# When Harry Met Peggy

# When Harry Met Peggy

## THE PERILS OF RESTORING A CLASSIC AUSTIN 1100

### HARRY TAYLOR

While the locations, the work carried out, the events and the time-line of the recovery work are all factual, apart from Harry and Marilyn, all other characters portrayed in Peggy's story are a combination of several people involved at that time. Any resemblance to any one individual is wholly coincidental.

Front cover illustration: Harry and Peggy (Author)
Back cover illustrations by David Augustine

First published 2026

The History Press
97 St George's Place, Cheltenham,
Gloucestershire, GL50 3QB
www.thehistorypress.co.uk



British Library Cataloguing in Publication Data.
A catalogue record for this book is available from the British Library.

ISBN 978 1 83705 053 6

Typesetting and origination by The History Press
Printed and bound in Great Britain by TJ Books, Padstow, Cornwall.

The History Press proudly supports
Trees for Life
www.treesforlife.org.uk

EU Authorised Representative: Easy Access System Europe
Mustamäe tee 50, 10621 Tallinn, Estonia
gpst.request@easproject.com

# Contents

# Acknowledgements

I am indebted to the following people for their support in getting this book published. They have only themselves to blame.

Firstly, thank you to The History Press for having faith and agreeing to publish the Peggy story, and to Amy Rigg and Jezz Palmer for all their help, advice and patience.

To my amazing brothers, John, Russell and Robert Taylor for your creative ideas and all the laughs along the way. A special thanks to Rob for collating the magazine articles and sending around to publishers.

To Dave Wilkins, editor of *Idle Chatter*, the official Austin 1100 club magazine, and a thoroughly nice bloke! Thank you for giving me the opportunity to write the articles for the club magazine, for your encouragement in making it into a book and for writing the introduction.

Special gratitude and thanks to David Augustine for his wonderfully witty and clever illustrations, each one a work of art.

To Peter Buller for keeping both me and Peggy on the road since 2023. Thanks mate, sorry about the gear box and clutch!

To my life partner Marilyn Timms for her unfailing support and understanding; she's been the voice of reason throughout. I just wish I'd listened more ...

And last but not least, to our cat, Athelstan AKA Stan. Thanks for causing chaos at every opportunity, for walking across the keyboard, sitting on printed pages when soaked by the rain and, in essence, for being a cat.

My dad had one of those …
The story is bigger than the car.
Having said that, it wasn't a particularly big car.

Watching out for old folks on the way to Sandwich, Kent. (Maddy McCoy)

# Foreword

Back in January 2020, I received an email from 1100 Club member Harry Taylor in my capacity as editor of *Idle Chatter*, the club's glossy bi-monthly publication. It contained the first instalment of the restoration story of Peggy, his Austin 1100 MK I Countryman. I am in the lucky position of being well supported in my editorial role, so I receive emails from members with magazine submissions pretty frequently, but quickly I realised that this one was different – a couple of paragraphs in I was already laughing, plus at the same time feeling a huge sense of empathy with the varying emotions that Harry was experiencing! I knew other readers would feel the same, so I immediately decided it was going to be published. I was already looking forward to the second chapter when suddenly the title of the series became crystal clear – it could only be 'When Harry Met Peggy'!

Two months later, the second instalment landed in my inbox and I couldn't wait to catch up with Harry and find out whether the extraction of the car was a success, even though I knew it was in the end because I'd seen pictures of the finished car! And that set the scene for what would ensue every two months from that point onwards, all through the Covid period and beyond, meeting the various dubious characters along the way (including the ubiquitous Barry) and often sharing a metaphorical beer with Harry when he ended up in the pub at the end of another demoralising turn of events. Progress with the restoration was slow, but progress there was, and eventually, with some sadness, we reached the end of the tale. A few times during our email interactions, Harry and I joked that there could be a book in it, but little did we know ... there really was! And I couldn't be more delighted to be writing the foreword for it now.

Of course, in our magazine, Harry was preaching to the converted – our club members are pretty much all as mad about the BMC 1100 range as he is! But what if you have picked up this book without that inbuilt predisposition? Well, you might remember 1100s and 1300s being around back in the day – there were around 2.2 million of them built from 1963 to 1974, and they were the best-selling cars for most of the 1960s, so nearly everyone owned one or had some experience of one. They were the brainchild of the inspirational Sir Alec Issigonis, also the creator of the one and only Mini, the 1100 being essentially a stretched version of that great little car. Then they were styled by the legend that was Pininfarina, and engineer Dr Alex Moulton made them float on fluid – no springs, no shock absorbers, just hydrolastic displacers and some shiny green liquid! Add to that an expanded A Series engine

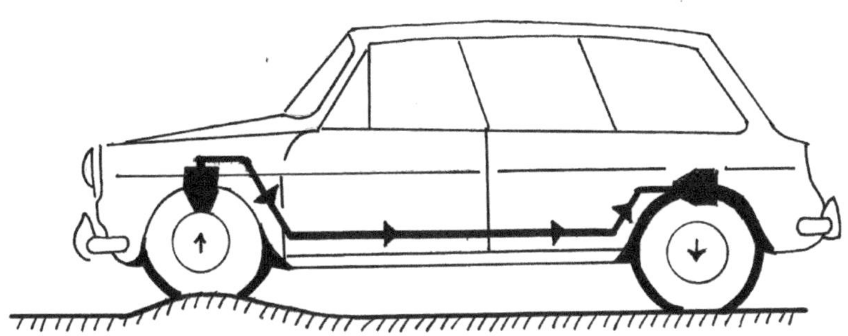

Hydrolastic.

mounted sideways, and we had a winner. A brilliant, versatile, comfortable daily workhorse. There were different badge options for loyal customers of the well-known marques of Austin, Morris, MG, Riley and Wolseley, and a luxury version trimmed by Vanden Plas. There was also a sporty twin-carbed GT model and, like Peggy, an estate with a lift-up tailgate. Truly something for everyone.

One slight problem – they all rusted. They rusted a lot. Many of them sadly didn't live very long for that reason. Nowadays, we are used to dealing with that, though. We have a great owners' club that provides lots of support, and we have people like Harry who are willing to take examples that should probably be scrapped and return them to their former glory. This is the story of how such a process can take over your life, exasperate your nearest and dearest, empty your bank account and lead you to meet a cross section of random and sometimes quite brilliant people along the way. I very much hope you enjoy the read as much as I did.

Dave Wilkins (Editor, *Idle Chatter* magazine)

# Preface



This is the story of a car and a man. There are other people in this story. There are even other cars. However, in its essence, this is a story about one particular car and one particular man. The power of the story lies somewhere between the vehicle and the human being who rescued it from a muddy field in Somerset and gave it life. We follow the relationship between Harry and his Austin 1100 Estate as he raises it from the dead and restores it to its former glory. This journey is a story of indelible memories, stubborn perseverance, hilarious mishaps, comical characters and frequent visits to the pub.

Readers, if you happen to be in East Kent and you see a Snowberry white 1967 Austin 1100 estate zipping around the narrow country lanes, give it a wave. That's Peggy, a car that has risen from the grave; a car with a story.

# GET AHEAD: GET AN AUSTIN 1100

**HEY YOU. *WHO, ME?* YES, YOU!**

Ever felt like your life was on autopilot? Are you trapped in the same humdrum routines every day? Is the most exciting thing in your life the weather, finding a matching pair of socks or some loose change down the back of the sofa?

It sounds to us like you're stuck in a rut. Well, don't worry, you've come to the right place. We know exactly what you need. No, not gold and jewels or a date with Brigitte Bardot. It's an Austin 1100.

Everybody's dream has to start somewhere — and some of the best have started in an 1100.

George Best in 1976. (Dutch National Archives via WikimediaCommons)

## EVER WANTED TO BE A MANCHESTER UNITED LEGEND?

Young George Best did, but he was struggling to do keepy-uppies in the Salford reserves — and that's where he might have stayed, had he not had the bright idea to get himself an Austin 1100. Sure enough, he passed his driving test as easily as he'd pass five or six defenders; then he made the United first team and won a European Cup, two league titles and a Ballon d'Or. All in a day's work for our friends at Austin. Well done, Sir Alec Issigonis!

**IT'S NOT ALWAYS EASY TO BECOME A MUSICAL SUPERSTAR AND CULTURAL PHENOMENON, BUT YOU CAN MAKE THINGS A LOT SIMPLER FOR YOURSELF BY GETTING AN AUSTIN 1100.**

John Lennon drove one in the early days, before the Beatles even had a recording contract. With plenty of room in the back for a couple of Rickenbackers, plus the entire line-up of Sgt Pepper's Lonely Hearts Club Band, success was assured. The Woodies never had an Austin in their back catalogue. Who are they? Exactly. Capital stuff, Battista Pininfarina!

**DO YOU DREAM OF A CAREER IN COMEDY? THERE'S STILL HOPE.**

Young Billy Connolly was a struggling stand-up in the late 1960s, getting fewer laughs than the Chancellor's Autumn Budget Statement. The only time people laughed at him was when he said he wanted to be famous. So he bought an Austin 1100 to get himself to gigs. Now Billy is an internationally celebrated comedian — and no one's laughing now, I can tell you. Congratulations to the lads at Longbridge!

**HAVE YOU HAD A LONGING TO BE KNIGHTED?**

It might be worth remembering that Sir Nikolaus Pevsner, the world-renowned historian, was once just plain old Nick, fond of a bit of reminiscing — until he got his Austin 1100. Suddenly, out came Queen Elizabeth's sword to dust off his shoulders. She knew a winner when she saw one. Well done Liz and hats off to the boys at Cowley!

As if that wasn't enough, Prince Phillip was seen behind the wheel of an 1100. We can't promise owning one will make you royalty, but it will make you feel like royalty, which is just as good (if you don't include the big house, yacht, country estates, crown jewels, untold riches, endless relatives, tiny dogs, etc.).

**HAVE YOU GOT AN ITCH TO BE AN EXPLOSIVE, VOLATILE HOTEL OWNER? LET'S FACE IT, HAVEN'T WE ALL.**
Well, you can scratch that itch by taking a stroll down to your local Austin dealership. Basil Fawlty was the highly regarded proprietor of a well-respected boarding establishment until he got himself an 1100. After that he became rude to guests, ruined the food, destroyed the hotel and couldn't tell a rat from a Siberian hamster. In short, he never looked back. But remember, don't mention the war. Let's hear it for the chaps at Morris Oxford!

Famous people who did not own an Austin 1100 include Genghis Khan, Vlad the Impaler, and Attila the Hun.

# DON'T BE AN ATTILA — BE A GEORGE BEST. GET AHEAD: GET AN AUSTIN 1100.

# Prologue: 1967

I'm going to take you back in time. Close your eyes. Forget about the meetings, appointments, the plans and arrangements. Relax. Breathe deeply and slowly.

Brace yourself. The journey is beginning. Leave behind your mobile phones, computers, robot hoovers, microwave ovens, air fryers, self-parking cars, virtual friends, and remote controls. Forget about modern conveniences. Where we're going, little is modern and even less is convenient.

Keep going back. The air is getting cleaner, the roads are quieter and you can swim in the rivers. Keep going back. The skies are almost empty now and there's snow in the wintertime and sometimes snow in spring and early summer too. Keep going back. People are talking to each other, goalposts are made of jumpers and you can get an appointment to see a doctor, even if there isn't much they can do for you.

Now open your eyes. You've landed back before the dawn of time – well, digital time, anyway. The year is 1967, a year unlike any before or since. This is where our story begins. Step outside, take a look around.

It was a time of Brendas, Kevins, Cynthias, Nigels, Trevors and Sandras. What a time to be alive.

The shackles of post-war living are being thrown off, though history suggests that it might have been better if some of these shackles had been left in place. Britain still had a manufacturing base, making ships, cars, iron and steel. The production of white goods was booming, with Hoover, Creda, Belling and Moffat making washing machines, vacuum cleaners and ovens. These things are made abroad now, undermining the economy and making the trip to return stuff a hell of a lot longer.

Manchester United had just won the league title, which was the key that would unlock their first European Cup win, and George Best had arrived in the shape of an Irish angel to deliver the country from two-footed tackles and provide Miss Worlds with someone to date. England was still basking in the glow of its World Cup victory the previous year, before seeming to decide that it had all been too much like hard work and subsequently putting all its energy into making Carry On films instead.

It was a time before satellite navigation, when finding your way anywhere involved unfolding a map the size of a cricket screen over the steering wheel while doing 70 in the fast lane. Either that or thumbing through an A-Z Guide where you had to cross reference grids and page numbers before sending the result off to GCHQ to decipher and tell you whether to make a U-turn or not. In spite of this, everyone arrived at their destinations without sat-navs, though they were all generally two days late.

Smart motorways had yet to be invented, and the hard shoulder was available for picnics, taking the dog for a walk, or brief caravan holidays. And roadside assistance was more about asking someone for a push than calling the AA or RAC.

It was the time of Flower Power, just before it was inevitably replaced by the more reliable nuclear power. Donovan's 'Sunshine Superman' had been a major hit, and tie and dye T-shirts and long hair were the compulsory uniform of those who didn't want to conform. It was the Summer of Love. Young people across the country insisted that everyone should love each other and wouldn't stop going on about it until everyone hated them for it. Eventually, they gave up and went on to smashing up bus shelters, which no one seemed to mind as long as they weren't going on about love.

Leonid Brezhnev was President of the USSR and Lyndon B. Johnson was President of the USA, while Harold Wilson smoked a pipe for Britain. In his gabardine mac, Wilson cut such an inconspicuous figure that he had served a full term as prime minister before anyone realised he'd been elected.

There was a new kind of war. It was called the Cold War, where East and West silently fumed at one another, as if they had each filled the dishwasher when it wasn't their turn. Instead of being fought with traditional armaments, Britain used public school boys. It turned out that they had more secrets than the country they were supposed to be spying for. Eventually, the public school boys gave everything up for a life in Moscow, in small apartments in brutalist blocks where they couldn't understand a word anyone was saying. It seemed an awful lot of effort to achieve a life they could have had by simply moving to Birmingham.

The safety of children was an emerging concern. A public information campaign was led, as you would expect, by a family of squirrels, squirrels being well known for their careful demeanour and non-risky behaviour, while hanging upside down by their toes. It was called the Tufty Club. The Club featured Tufty the Squirrel and Willy the Weasel. Tufty always did as he was told and never got into trouble, whereas Willy did not do as he was told and learned the hard way, by getting run over. The main lesson children of the day learned from this was not to be a weasel.

Morals were now seen as something you had to guard, as if someone might melt them down and sell them abroad if you weren't looking after them. You couldn't go into a cinema to see a film that contained swearing unless you were 18, although you could play in a cement mixer when you were 8, as long as you didn't get sand in your plimsolls. It was enough to drive children to sweet cigarettes.

The post-war leisure boom was in full swing, and people had a greater disposable income, though this wasn't hard as they'd had absolutely no money at all in the years leading up to 1967. People had more money, but nowhere to go. At this time, they were still learning to speak loudly to foreigners so they could make themselves understood and then go abroad on package holidays. While they were waiting, they bought caravans instead. Essentially, these were mobile cage fighting arenas where you packed your family into a small space and thrashed out all your grievances in a week near the sea at Brean, returning home with whoever had survived, like an early version of the Hunger Games for the family.

The space race was in full tilt. NASA was developing the LVR – Lunar Roving Vehicle. It turned out to be marvellous for exploring the Moon, but had relatively little boot space, certainly when compared to something like the Austin 1800, and so was completely impractical to take on a Sunday shop. Now, I'm not saying that the 1800 is better than the Lunar Roving Vehicle, but I will say that a quick glance at the classic car classifieds shows there are around thirty 1800s currently up for sale, while there are no LVRs. Facts are facts I'm afraid, NASA.

The Beatles released the iconic *Sgt Pepper's Lonely Hearts Club Band*, which was widely acclaimed by critics as a work of genius and was Britain's biggest-selling album of all time until the release of the less genius, *Now That's What I Call Music 25*. The Rolling Stones produced *Between the Buttons*, which included the classic singles 'Ruby Tuesday' and 'Yesterday's Papers'. However, some of their other hits that year saw them arrested for drug possession. In 1967, the Beatles were not long for this world. The Rolling Stones had already decided they were going to outlast the world.

It was the best of times, it was the worst of times, it was the *TV Times*. BBC 2 became the first channel in Europe to begin colour broadcasts. You have to remember that this was at a time when even some of the rainbows were still black and white. Programmes like *The Avengers*, *Coronation Street* and *Morecambe and Wise* ruled the ratings. Onto this landscape, a show called *The Prisoner* was launched. The series followed an unnamed British secret agent (referred to as Number Six) who resigns from his job for unknown reasons. After his resignation, he is mysteriously abducted and wakes up in a bizarre, isolated place known only as 'The Village'. The series' popularity was mainly down to people wanting to resign from their jobs for unknown reasons.

And there were cars. Yes, I know we have cars now and we had cars before that, but in 1967 there were *cars*. The journey they took us on was life – the Hillman Imp and the Mini for those who were young and single, the Vauxhall Viva, the Austin 1100 and the Ford Cortina MK2 for when you'd acquired a family, and then, when you'd got that promotion you were after, there was the Rover P5 or the Ford Zephyr or the Jaguar MK2. Finally, when you'd retired and the children had left home, you could at last get that sports car you'd always wanted, perhaps an MGB, a Triumph Spitfire or, if you'd done really well and your kids had moved a very long way away, an E-Type Jaguar. These cars were our lives and you could not separate us, and you still can't to this day. They sprang from the imagination of great men like Sir Alec Issigonis, Roy Haynes, William Lyons, David Jones and Syd Enever. Men who could see into our hearts and designed what they found there.

While all this was happening, in a nondescript street in a nondescript suburb of Birmingham, an 8-year-old boy called Harry Taylor was juggling a leather caser as well as he could do with something that weighed more than he

Harry, 8, in his Villa kit. Already displaying a love of lost causes.

did. Harry was a bony kid with a home-made haircut, skin the colour of skimmed milk and knees like coconuts. He lived in a two-bedroom council house where the only modern convenience was a roof. Being the oldest of four brothers meant he was never alone, but also that he only got a quarter of the food he'd had before the others arrived.

Around the time Harry was learning that you couldn't do PE with any dignity when you kept forgetting your shorts, just a few miles away, a mineral blue Austin 1100 was rolling off the production line at Longbridge. To the hum of the factory floor, someone, probably a Brian, installed the badge, the final touch that transformed the car from a collection of parts to an icon of British motoring. Someone else, most likely a Keith, made a final inspection with clipboard in hand, before leaning over to remove a speck of dust and then smiling proudly as she rolled out of the factory to await her test drive. The world would never be the same again. Well, at least not Harry's world.

The Frank Capra gods were soon to take control and bring them together.

Close your eyes again. We're going forward this time. Harry Taylor is on a date with destiny.

1

# Snowberry White, with a Black Dual Strip

Have you ever owned a car that has somehow stayed with you long after you parted company? Even the smell, the sound, lingered in your mind?

Now, there is always the risk of anthropomorphising mechanical things, but this was never the case with my first car. It was a machine for performing tasks and really that was all I have ever thought about any car I have owned. Yes, some were nicer than others, and more worthy of conversation, while, honestly, many were just plain embarrassing. However, they have all left me with many barroom stories that I reach for with alarming regularity while gazing into false smiles that resemble grimaces and glazed eyes that I recognise even through the heaviest of lenses.

My first car was a MK2 1969 Austin 1100 Estate. Snowberry white with a black dual stripe and funky square

Even then the world was leaving me behind; it was 1977.

plates, no wheel caps and a horn that loudly and jauntily played the Mexican Hat Dance as it drove around the streets of Erdington, Birmingham. Under the bonnet lived a 1275 engine with a single carburettor. Seated behind the wheel was a cool dude clad in denim flares and a Francis Rossi waistcoat. Even then, the world was leaving me behind. It was 1977.

This particular motor had been sold to me by a friend of my father who was a mechanic. It set me back £180. It had the usual issues even then. The inner front wheel arches and floor were thickly covered in layers of underseal. This disguised a creative combination of welding and fibreglass. It gave the car the distinctive smell that has stayed with me to this day. It used to eat front wheel bearings (due to CV joint issues which I did not understand at the time) but boy, it was fun, as long as you understood the enigmatic gearbox.

In the end, it was that gearbox that sadly spelt the end of our time together. I called a garage to ask about either getting it mended or having the whole box replaced. The conversation went like this:

**Me**: Hello, I have an Austin 1100 with a gearbox problem. Can you help me?
**Garage**: (sharp intake of breath) *1100 – or 1300 engine?*
**Me**: *1300.*
**Garage**: (another sharp intake of breath) *What year?*
**Me**: *'69*
**Garage**: (an even sharper intake of breath) *Rod change?*
**Me**: (getting a little worried) *Dunno …*
**Garage**: (a short silence which doesn't help my growing worry, followed by a further sharp intake of breath which only adds to my anxiety) *Two door or four door?*
**Me**: Two door.
**Garage**: (some tutting, followed up, of course, with a sharp intake of breath) *I see.*

This conversation went on like this for some time. I think I gave up the enquiry when we had reached the stage of describing the exact style and colour of the door handles. I was unable to endure one more tut or an even sharper and more prolonged intake of breath.

It was the end. In fairness, the car had carried me, my Mom and Dad and my three brothers around to quite a few jumble sales and the odd memorable day out to Warwick. However, in the cold Winter of Discontent of 1979, OOG 300G departed Dean Road, Erdington, for the last time, chained up on the back of a hulking scrap wagon.

In the intervening years, I had never given the car much thought. It only cropped up in small snippets of conversation with my brothers when we met up as relatively fully formed adults.

Fast forward to 2010. My long-suffering partner Marilyn and I moved to a cottage located in a conservation area in the London borough of Bromley. Of course, this all sounds very nice, and indeed it should have been. Here's a non-car-related lesson for you lucky readers: never buy a listed property that has bizarre alterations that cannot be explained by the slightly reticent and barely sober surveyor. Especially if it is marketed by the estate agent using the phrase 'interesting opportunity'.

During the trials and upheaval of having floors replaced, ceilings torn down and our entire kitchen removed, we took to escaping the chaos at weekends by relaxing in a local coffee house and reading newspapers and magazines.

It was on one such occasion that I picked up a copy of a classic car newspaper. Thumbing idly through the sales adverts, I noticed an Austin 1100 Estate in Snowberry white for sale for £4,000. I showed it to Marilyn and explained that one of these was my first car. She quickly glazed over; her natural defence in these situations. But in fairness, she did say that perhaps I should get it if it was what I really wanted. I pondered for about thirty seconds (one of my longest ponders). I decided to put the notion away. How frivolous, I thought.

The moment did sow a seed in my mind, however. Some months later, the house became liveable, to the extent that I could actually find a space to sit down and look at things on the internet. One Sunday, that seed in my mind began to grow shoots. I typed in 'Austin 1100 Estate'. An entry on eBay flashed up. I looked. I looked again. Was that really an Austin 1100 Estate? Well, the rear end certainly looked familiar but I couldn't tell about the rest, as a large proportion was buried in a field in Somerset where it had apparently lain for some time. I later found out that it was last taxed in 2004.

I looked at the price ... £500. I tried to think rationally. I wanted one but was this it? After all, it was buried in a field and it certainly was not going to be driven away. The advertiser didn't even know if it would run. No! This is way too risky. Final decision. I was sure.

As I relaxed and sat back from the screen, my hand, under no conscious guidance from my brain, shot forth and with unerring accuracy slammed a firm finger down on the bid button. I wrestled with my keyboard desperately trying to withdraw the bid, but it wasn't happening. I was scared. Then I thought: there is bound to be some other person out there who understands how rare an early 1967 MK1 1100 estate is. After all, it is a bit of British heritage and there are loads of committed and wealthy old car enthusiasts out there who will want this. Furthermore, there were still two weeks left to bid. I relaxed. My world wasn't ending after all. Things would work themselves out.

2

# How Hard Could it Be?

Just over two weeks later, I received a notification email that informed me I needed to pay for my newly acquired vehicle. I panicked. I called a work colleague who knew a bit about eBay and she suggested that I send an email informing them that there had been a horrible misunderstanding.

My work colleague's helpful suggestions were:

Misunderstanding 1: I have a mischievous child who has hacked my computer and put bids on random, ridiculous items.

Misunderstanding 2: My overactive cat ran across the keyboard of my laptop and accidentally placed a bid on the item.

Misunderstanding 3: I was extremely drunk and fell asleep on my computer.

Misunderstanding 4: I suffer from the recognised medical affliction: 'sleep typing'.

I thanked my colleague, but as I read through the 'excuses', I found them more and more unbelievable. I am also a terrible liar. I just had to come clean. I talked to Marilyn and after much mirth, she said, 'Go for it!' I pondered again – of course, I should go for it, it is a challenge and coming through a challenge like this would be character building. I'd seen some Walt Disney films so I knew about these things. At times it would be difficult but I would ultimately triumph, becoming a better person in the process and making the world a better place. I now know why Disney has never made a film about restoring an early '67 MK1 1100 estate.

In the next few days, I set to the task. I contacted the vendors and asked a friend if I could situate the vehicle on his driveway for the short time it would take to get the car back on the road. I mean, how hard could it be? I then undertook the task of arranging the transport for my glorious acquisition. There was a site that magically set all the providers to bid against each other for my work and, eventually, I got a price I was happy with. This was going well – project management at its finest.

Dennis, who had donated the driveway, also offered to come with me to the farmer's field where my buried treasure lay. We set off on a glorious day in early spring, flying down the motorway, exchanging jokes and anecdotes. You know how blokes do. Coming off the motorway, the 'A' roads got smaller and narrower. The 'B' roads got even smaller and even narrower until we were travelling on

Then, there it was.

something that would have been considered a bit dodgy even in a Thomas Hardy novel.

After two and a half hours, we arrived at the picturesque pub that stood near the gate leading into the field. The sun was shining, throwing beams of optimism on the wonderful rural scene. An old man duly wandered out of the pub and spoke in a minimal, monosyllabic dialogue. I felt like I was talking to the farmer in the film *Withnail and I*. With a combination of short words and hand gestures, he indicated that we should follow him into the nearby field. This turned out to be the *first* field. We were quickly led into *another* field.

Pretty soon, as we entered the top gate of the *third* field, I came to the realisation that this particular area of rural England had been dealing with a lot of heavy rain very recently. My feet sank with every step. My heart sank with every other step. The mud that came over the top of my newly acquired shoes was now adding unwanted colour to my socks.

Then, there it was. It was still a way off (probably at least another field) but there was no mistaking it. Scratched and faded blue paint daubed with bird poo could not camouflage the familiar shape. As I got closer, I remember thinking: I'm sure they used to be taller than this. It was at that stage that I realised the car was either very deeply buried in the field or the bottom half of it was missing.

3

# Almost Without Realising, I Had Started to Care

Still thinking of getting out of the predicament I found myself in, I was deep in thought, as well as deep in mud. In the back of my mind (which is a very dark place where many embarrassing things are stored) I had formed a cunning fallback plan: 'Oh, look at that, the car transport wagon has not turned up!' I had even rehearsed being unable to find the contractor's mobile number.

Alas, a rumbling sound announced the ruination of my plan. The wagon appeared, large and white, at the top of the last field, trundling towards us, getting larger and whiter by the second. Still, perhaps the ground was too soft and the driver would scratch his head and say: 'Sorry, can't get near it, mate.' This notion quickly disappeared as the truck, scornful of the muddy conditions, made its way alongside the car much like a prima ballerina pirouetting her way neatly to the centre of her stage. I picked my way through the muddy ruts to the driver, who seemed reassuringly unimpressed with the car, and with me.

We talked back and forth about the best way to load the car. We decided to attach the winch wire to the front subframe of the vehicle. This necessitated the car being turned around in the boggy field. As the car was painfully turning, inch by inch, I looked over to where my friend Dennis and his dog were standing. Dennis had decided to chat up the farmer's wife, who had strolled over to see the spectacle. I watched as his complete lack of social skills gradually drilled her into the ground. Dennis's dog, a friendly mongrel who was along for the ride, was doing much better and had started enthusiastically humping her leg. If any dangerous liaisons were to be achieved, my money was firmly on the dog. I looked further round the field and my sight fell on the farmer. He was looking away, squinting at something far in the distance. The whole scene was surreal.

I was shaken out of this episode of *The Twilight Zone* by a shout from the transporter driver, who had now moved on to winching the car onto the back of the truck. As this was happening, I was suddenly struck by the possibility that due to the lack of a floor pan, inner wings and sills, the rear of the car may decide not to follow the front of the car onto the carrying bed of the transporter.

'It's not going to go on!' the driver shouted. I hurried over, about to explain my thoughts about the structural integrity of the car, when the driver anticipated my concerns. He turned towards me and let out a familiar sigh. He pointed out that due to the fact that the suspension was on the bump rubbers and the tyres were flat, his opinion was that we might damage the exhaust if the winching continued.

I rose to my full height of 5 feet 8 inches, and stood right in the driver's face. Quite forcibly, and in my best impersonation of the man who did the voiceovers for movie trailers, I said, 'It's going on!' Almost without realising, I had started to care about this thing stuck in a muddy field in the middle of nowhere.

The driver looked impressed, pulled himself together and carried on the winching, albeit at a snail's pace. Halfway through, sure enough, there was a dreadful crunching of metal and a scraping sound that reminded me of a distressed whale song. The farmer's wife turned round, wide-eyed, Dennis's dog stopped humping in surprise and even the farmer himself looked over with a puzzled frown on his ruddy face. Dennis, supremely oblivious, carried on telling his fascinating life story to the farmer's wife, only finishing when the car had finally squealed its way to a comfortable berth on the transporter and the ensuing silence made him finally aware of his own voice.

So, there it was. The driver confirmed the address where the car was to be dropped off and the transporter waddled its way out of the field and back onto the road. I reached into my pocket and marched over to where the farmer had stood for the duration of the performance. I fumbled for the roll of notes; £500. A lot of money.

'So, how much do I owe you?' I asked hopefully. I was expecting a show of empathy, a recognition that a fellow human being was massively out of his depth and had just signed himself up to five years of heartache and disappointment that would rob him of every spare hour he

Dennis, supremely oblivious.

had and an amount of money to match. Our eyes met, there was a flicker, a look down, a look up and a look at Dennis's dog, now happily re-humping the farmer's wife's leg.

'Five hundred it is then,' he said, smiling. He counted the money in double quick time and stepped assuredly away across the fields and back to his farmhouse. I rallied Dennis and his dog and we made our way back up through the fields to Dennis's car.

The ride back was as unremarkable as only the M4 and M25 can be. We arrived back in Beckenham just as the transporter with the car rumbled up behind us. The scarred and mud-spattered car was duly lowered onto the forecourt of an old house surrounded by high hedges and towering trees. This was good, as there was no garage to hide my acquisition in. I paid the transporter driver from my rapidly diminishing roll of notes.

There we were; the car, Dennis, his dog and me. 'Right,' I thought, 'this needs decisive and careful planning if the vehicle is going to be back on the road in six months. I will need to source parts, find expert labour and timeline everything to perfection.' It could be done! I felt a surge of optimism.

I leant on the roof of the car and immediately jarred back from the sharp pain in my elbow. I hadn't noticed that the roof was completely and jaggedly rusty. Yes, even the roof was rotten. Some misguided brain transplant donor had decided to crown the car with a mock vinyl roof – a vinyl roof on an estate car? In realising their failure, they had ripped off whatever unholy material they had used as a covering. As the powerful adhesive had melted itself into the paintwork of the roof, the removal of the covering had also duly removed any remaining paint. Ten years of rain, snow and sun had done the rest.

All of a sudden, I was transported into the scene from the film *The Killing Fields* when the character escaping the Khmer Rouge is wading across the paddy fields. He falls into the watery field and realises with horror that he is walking on the bones of the dead. In the driveway, I imagined the camera pulling back to a high, wide shot and the strains of Barber's 'Adagio for Strings' becoming louder. My feeling of optimism took one look at this scene and swiftly removed itself from the driveway.

At least, I reflected, I had one option that the movie character did not.

I looked at Dennis. 'Pub then?' I said.

4

# This Could Be the Start of a Beautiful Friendship

I decided to name the car 'Peggy'. A quick glance at the licence plate should help you figure out why. The following days after she had landed in Beckenham from the distant climes of Somerset were relatively uneventful. Most of the week had been spent dealing with the inevitably thorny issues that working in social services generally provides. However, underneath it all, I was formulating a plan. To be honest, the intervening years have now led me to believe that it wasn't a very good plan. In fact, the evidence in retrospect is overwhelming, as you will read in my further recollections.

By Thursday of that week, I was simply unable to keep myself away from the piece of ground where Peggy was parked. Sat down squat on the suspension rubbers, she looked little like the car I remembered during the heady days of the '70s. It was low, long and the sharp tail fins looked like something from another century, which, of course, they were.

Right, I thought, let's jack it up and have a good look at the underneath. I had a powerful 2-tonne trolley jack and four axle stands. I should have known that very afternoon that if jacking the car up was not easy, the rest of my plan might not be entirely plain sailing either. I had to use a very small scissor jack under the front subframe to prop the car up high enough to push the trolley jack underneath. I managed to angle it to a place on the subframe so I could withdraw the scissor jack and carefully replace it with an axle stand. This awkward task took some time but, undaunted, I finally raised all four corners of the car onto axle stands. The car was set firmly just above the natural ride height. I wiped my hands on the back of my trousers and stood back. I then got a sense of just how good these cars could look. Doubt had finally been overcome in a titanic struggle, to leave trepidation as the victor.

If you have ever hurt yourself in an ill-thought-out DIY project or in an overly aggressive amateur football match, and it has been one of those injuries you just cannot bring yourself to look at, then you will understand how brave I was in putting my head underneath the car. It wasn't the fear of the vehicle falling on me; my head was far more robust than the shaky vehicle currently suspended above me. In truth, I was afraid of what I might find lurking beneath the car.

As I gathered my courage and gradually looked up, it occurred to me that, as children, we spend a lot of our time being scared of the dark. Now here I was as an adult, feeling very scared of the daylight and what it might reveal. I knew exactly how Christopher Lee felt, I can tell you!

There wasn't actually that much daylight at first, but there were a few strange shapes on the inner wings on both sides that didn't appear to match anything that Pressed Steel Fisher had turned out. However, the gentle curve of the inner wing over the steering rack and drive shafts seemed remarkably true in shape. Of course, that was before I naively wielded the obligatory heavy screwdriver. Large screwdrivers wielded underneath older cars have the destructive power of a swinging lightsaber. So it was with the scarred and spattered screwdriver I had in my hand. A screwdriver that had previously only been considered good enough to stir large cans of paint was now tearing ragged lumps from the underneath of Peggy. It wasn't pretty. With every jagged hole came the vision of Captain Smith standing on the bridge of the *Titanic* being told by Thomas 'Screwdriver' Andrews that the holes in the hull were a little more concerning than they had at first thought. I flailed my screwdriver back and forth in the near darkness, hoping to strike something solid. I knew the car was old, but even Howard Carter had found some metal worth restoring in the first weeks of his Egyptian explorations.

Breathing deeply, I lay back with my head on the hard driveway. This was bad, very bad. My blackened face squinted into the daylight where the junction of the floor and toe board met the sills on both corners. At least this meant I didn't need the doors to work – I could always gain entry via these holes. The imprint of the floor pan on both sides was just that; an imprinted pattern preserved

At least this meant I didn't need the doors to work.

by grass and mud and cardboard and glue but alas, no metal. Still, it could be worse. The inner wing was still there and seemed quite hard to the touch. One strike with the lightsaber-screwdriver on the inner wing revealed a strange truth. At some stage, the inner wings appeared to have been repaired by either Valerie Singleton or Tony Hart. These were the only two people I was aware of who had the skills to manipulate papier-mâché in such a way. They must have, at some stage, sold their souls to the evil first owner of the car and were beholden to do his bidding or be revealed to their respective children's clubs.

By now I was a little concerned about the destructive power of my screwdriver-lightsaber and for my safety underneath the vehicle. I scrambled and scraped my way out and stood up. I dusted myself down and fixed my gaze on the windscreen where the last tax disc was still clinging on grimly. It read '2004'. It was now 2010. I assumed that the previous owner must have given up attempting to get the vehicle MOT tested at the same time that Ray Charles and Stevie Wonder had closed down their testing station business in the area.

Remember, readers, how I told you I was formulating a plan? I think it was on the following Wednesday after the inspection/autopsy when I had my eureka moment. The government had just announced the collapse of everything that needed money. Where there is a crisis there is an opportunity, thought I. An opportunity to get my car repaired and also to do some social good. So, in my lunch break, I set off to the local newsagent with a freshly typed advert, which I placed for two weeks. Two weeks should be ample time to set my plan in motion. I felt like Peter Lorre about to embark on some underhanded deed.

I will let you into my reasoning, which you might file under 'It seemed a good idea at the time'. If people were losing their jobs and in need of an income (which they were) it stood to reason that amongst those in this predicament, there may well be someone who was both a good mechanic and also handy with a welding torch. OK, OK, I can hear your laughter … yes, yes, OK. Stop right now.

But it worked! (So there!) I received a phone call from a guy just two days later. It was a man called Barry who told me everything about my car, from the gear ratios to the paint colour number. I was stunned. In the words of Humphrey Bogart, I thought, 'This could be the start of a beautiful friendship.' I arranged to meet up with Barry one lunchtime. It was a meeting that would change both of our lives over a period of five years in ways we could not have foreseen. At that moment, my optimism had returned. Hope had stirred. I needed to do some more thinking – pub!

5

# The Smell of Melted Car Radio

If you had been to Bromley High Street in or around 2010, you may recall that, apart from its rather swanky pedestrianised main street, on certain days it boasted kiddies' rides like a carousel or some sort of helter-skelter.

It was on such a day that I scuttled out from work and marched briskly the half mile to the high street from my office. This was the day I was going to meet Barry. Barry was the man who I was pinning all my future hopes upon. Without Barry, I was lost, and if I was lost, Peggy was lost too. Marching through the pedestrianised high street, I ignored the sound of banks crashing all around me and entered the area where the children's fun fair was whirling and swirling around.

Of course, I had no idea what Barry looked like. This was a typical bloke meeting – a five-minute phone call and a time and place is all that is needed, isn't it? No wittering on the phone about unimportant stuff like 'How the hell will I know who you are?' In fairness, Barry had described himself as working in the high street near the newsagents and having ginger hair. In bloke world, I figured that was as good as a DNA match.

I stood in the high street and looked around. The melodies from the merry-go-round and the carousel were clashing and morphing in my mind into the soundtrack from a Hitchcock film. All of the faces I could see were either blurred 10-year-olds distorted by the G force of the ride they were clinging onto, or moms looking oblivious or shouting, 'Keep hanging on Felicity, it will stop soon, I promise!'

My head began to spin too. Was it the beer I had the night before or the *lack* of beer from the night before? I wasn't sure. Then suddenly, amongst all the impossible shapes and noises, my eyes focused on a rather rotund chap with ginger hair. That had to be Barry – who else could it be? I started to move towards him, then quickly realised that this meeting was going to be more difficult than I had first realised. Barry was sat with a money bag on his shoulder right in the middle of the large carousel with jammy-faced kids whizzing around him at speeds that Apollo astronauts might have been reluctant to undertake in training.

When I was within 10 feet, I shouted to him, 'Barry! Barry!' Nothing. I moved closer and shouted again, even louder, 'BARRY!' Still nothing. I edged nearer and I could now feel the gaze of the mothers swing round from their blurred flying children and focus directly on the strange man who was shouting at the operator. I pushed closer and got to within a foot of the carousel. I was now in danger of becoming a human Van de Graaff generator as the speeding host of little nylon-clad bodies flew past me.

I leant in and shouted again, 'Barry!' Barry, for it was indeed my ginger contact, looked up. Our eyes met. Just at that moment, I slipped and landed on the hard shiny plastic of Billy the train, who was fortunately devoid of passengers. Off I went, joining the live tumble dryer that was the carousel. As I zipped past for the third time, I managed to introduce myself. Communication at this speed was indeed difficult and prone to the distortion of the Doppler effect. When Barry realised who I was, he heaved the rather large and theatrical brake lever. This had the effect of dispatching all of the children in a chaotic heap in front of their mothers. Barry then announced to the crowd

'Barry! Barry! Barry! Barry! **Barry!**'

that he was taking a tea break. We stepped down from the carousel and adjourned to a local coffee house, leaving behind the disbelieving faces of moms and children and all those who witnessed the unusual spectacle.

As we sat over steaming mugs of coffee, Barry showed me pictures of a 1968 Mini he was restoring. It looked great. 'It isn't quite there yet, but not far off,' he assured me. I filled Barry in on how bad I thought the floor and inner wings were on Peggy. 'Can you get the panels?' he asked. I told him that I thought I could get the floor but wasn't sure about the inner wings. Barry told me not to worry too much as there wasn't any panel he couldn't fashion from sheet metal. He then showed me pictures of working model fairground rides he had made. They were spectacular. We arranged an on-site assessment visit of my car. I left Barry to go back to the bemused crowds at the carousel while I ducked down a side road and took the long way back to the office, avoiding the incriminating gazes of moms and children, and possibly the photographer of the local paper.

That weekend, I visited Peggy, armed with petrol, a battery, some plugs and a few spanners. I gauged that a battery that ran my Nissan Micra 1.1 would do nicely for Peggy. I opened the bonnet and it all seemed so familiar, if a little dirty. I put Peggy in neutral and removed the plugs. I inserted a little oil and yanked on the fan blade and water pump pulley. The engine turned. There were no obvious water leaks or major oil leaks. The oil in the sump did look like Guinness, but it was there. I carefully gapped the new plugs and placed them in the engine, attaching the leads and putting the battery in the tray where it fitted snugly.

I adapted the fittings to make a decent contact on the battery and withdrew to sit in the car.

I was nervous. Wait! What about the choke? The cable was shredded but I trapped the pivot back to open using a piece of twig. I hastily returned to the driver's seat. Tentatively, I inserted the keys. All of the correct dashboard lights winked on. Wow! With a deep breath, I turned the key to engage the starter motor and the engine turned and fired but did not start. Even more wow! I did it again and the engine briefly coughed into life but died again. I was not disappointed. On the contrary, I was elated. This engine will go! I leapt out of the car to look at the engine, as I was pretty sure it was original with 37,000 on the clock. I jumped back into the car, turning her over again. The engine started and ran for about a minute. Amazing! I looked at the dashboard and noticed that it held a '60s radio in situ. If the engine had started, what else was still working?

It was, of course, at this point that I should have stopped. The day had exceeded my expectations by miles and I could have happily trotted off for a pint of Old and Filthy at my local with a Cheshire Cat grin all over my face for the rest of the evening. But no, I had to go and turn the radio on. To my surprise, the dial lit up exactly as it should have done. To my further surprise, the rest of the radio lit up too, and then duly melted. Even after a scrub and some fresh clothes, the smell of melted car radio stays with you, even in the pub.

I reflected on how I had actually been quite lucky. Many of you will have guessed that the reason the vehicle only ran for a short while and that the radio melted was because

I had put the battery on the wrong way round. I had forgotten that Peggy was earthed differently from modern cars. So there were more positives than I had thought … literally. I went to the bar for another pint and the barmaid came over. She theatrically stayed back while stretching for my glass. 'You smell of melted car radio, you do,' she offered with a grin.

'I know,' I replied, and thought about explaining, but it wasn't worth it. I was just hoping she hadn't heard about the man who climbed onto the kiddies' carousel in Bromley the other day.

6

# In a Galaxy Far, Far Away

You will recall in the last chapter, I had managed to meet the legendary Barry and arranged for him to give Peggy a look over. In the week that followed our first meeting, I had purchased a car cover, as Peggy was open to the elements on my friend Dennis's driveway. I got away that weekend to have a go at a few jobs I thought I could do myself. I parked up and walked over to Peggy and whipped her 'shroud' away with relish. If you have ever watched the Tony Hancock film *The Rebel*, you will know that he harboured his life's work in his rooms rented from Mrs Cravat. His masterpiece was covered under a similar shroud. He would pull the cover down from the hideous sculpture he had made, laughing with mock horror movie glee and shouting, 'There you are, me beauty!'

Well, unveiling Peggy felt a bit like that. She was a guilty secret and certainly no beauty at that stage. Although I did say under my breath in my best Hancock accent, 'There you are me beauty!' and burst into Baron Von Frankenstein laughter. I looked quickly round in case anyone at the bowls club next door had heard me.

By using some petrol from a can, I had managed to determine why Peggy would only run for a couple of minutes. I filled the float chamber and she ran until it emptied. I then searched out the electric pump in the boot and took off the hose to the engine but still nothing. I resisted the temptation to either bang the pump around a bit or to try to take it apart. I figured that it was the original pump and that even if it sparked into life, it would not be a life everlasting. Fifty years and a partial burial in a damp farmer's field would have taken their toll.

So, a pump was purchased, fitted and it duly obliged. Peggy was running! Better than that, Peggy was running well! I watched the temperature gauge reach normal and waited for it to go beyond. To my surprise and joy, it stayed just below normal. I revved Peggy and watched nervously for smoke in the interior mirror. There was none. I revved again and again … no smoke. I got out and walked to the rear … no smoke *and* the engine sounded good. I got back in and thought, 'Dare I try and select some gears?' At that stage, Peggy was on axle stands so there was no danger. I put my left foot down on the clutch but there was not one bit of movement. Looking under the bonnet, I could see that the clutch actuating arm was firmly jammed in position. I jarred it about a bit with the trolley jack handle but it would not budge. 'Oh, well,' I thought. 'At least there has been some progress.' I had bargained for a replacement engine and box so this was a real bonus. Incidentally, the engine and box were original and still trundle Peggy around to this day, sometimes at upwards of 65mph!

The next weekend, Barry arrived. He didn't arrive at the duly appointed time but he did arrive. Little did I know that Barry was setting a pattern that would become increasingly familiar. Barry owned a Toyota van, which I came to learn was more of a time-travelling machine than it was any formal type of four-wheeled transport. Barry's van had abilities only in certain aspects of time. The really advantageous aspects, like arriving at a place either before you have left or at least at the correct hour, were not factored into its operations. Barry would phone me on some days, telling me that he had just set off from his home (some 10 miles away) and would be on time for our rendezvous. This would mean that his vehicle would cover the 10-mile distance across south London in one minute fifteen seconds. Now that may sound amazing, but what was more amazing was that it actually took Barry three hours to travel the 10-mile journey. Somehow, he had disappeared into a wormhole and emerged from the other side in a galaxy far, far away. Three hours away, in fact.

Barry's journeys to work on Peggy were often punctuated with feats of seemingly impossible physics, mostly involving the expansion of time. I finally worked out that calls starting with 'I'm only ten minutes away' were often received while Barry's vehicle must have been travelling backwards towards its starting point. We have all experienced the arrival indicator board on train station platforms that shows the train to be two minutes late when only two minutes away. The board then changes its read-out two minutes later to say that the train is now ten minutes late.

The only possible explanation for this would be that the train at some point started travelling backwards or stopped completely. The delay in Barry's arrival times was made more apparent by the now miserable winter weather. It gave me much time to contemplate that when you engage an unemployed welder or mechanic, no matter how good they are, there is generally a good reason as to why they are unemployed.

Once he finally arrived, Barry took a long look at Peggy and seemed undaunted. He should have been daunted – very daunted. We settled on an approach that went a bit like this: now we have cut that bit out we might as well cut the rest out. So, it came to pass that we decided the front floor on both sides had to come out. I had the number for a supplier. Readers are probably aware of the great availability of panels nowadays. In 2010, the reality was somewhat different. The catalogue reassuringly showed every panel for every variant of the ADO16.

Filled with confidence, I called and asked for two front floor panels. There were what I can only describe as 'muffled scrabbling noises' on the other end of the phone.

A voice came back, 'We've got the offside floor front to back.'

'No front nearside floor?' I asked.

'No, I told you – we've got the offside floor – front to back.'

'When will you get the nearside floor in?' I chirped in a naively expectant Oliver Twist voice.

'I told you, we got the offside floor front to back – do you not know what you want?'

This wasn't going well – I needed to think. On the other hand, thinking really hadn't done me any real good so far

He should have been daunted, very daunted.

in this project. I knew that if I didn't get any panels, Barry would turn up with nothing to do and would expect to be paid. This would eat into my budget and also, importantly, my auxiliary beer fund. So, I ordered the panel – offside front to back.

One afternoon, I got off work early and picked the panel up from the Croydon Post Office delivery centre, which was situated on an industrial estate that must have doubled up at one stage for Moon Base Alpha. I gave my ticket to the sloth-like fellow behind the counter and he shuffled off through a set of doors from which I feared he would never have the energy to return. Return he did, though, with a huge cardboard shape that looked like it was concealing a piece of an early satellite. I signed and then we began a clumsy dance trying to get the panel under the counter gap or over the counter screen without taking out the light defusers. As I finally managed to wrestle the panel over the screen, the guy asked me what it was. I said, 'It's a piece of a satellite.' He nodded knowingly.

I took the panel to the driveway where Peggy stood, and on opening the doors, made a quick calculation. Well, truthfully, it didn't need to be all that quick as Barry was still 'on his way'. It is at this point that I will ask the forgiveness of many readers. Faced with one panel and two holes, I worked out that we could get enough patterned metal out of the panel we had to fit both front floor panels – just. And so it was that Barry welded the chopped in half panel into both front footwells. To this day, you have to look closely to see what we did, but it worked. Well done Barry – the ultimate reverse Time Lord!

7

# She'll Have to Go

Do you remember those wonderful Laurel and Hardy two-reelers that open up with joyous music? ('Dash and Dot', 'On to the Show' etc.). With the camera looking front on at the car as the pair are rattling along in a Model T? Of course you do. You may even remember this quote from Ollie: 'For the first time in our lives, we're getting some-place.' Well, that was how I felt about the restoration of Peggy in early 2010. It was going well. Slowly, but well. Barry was making appearances with the regularity of Van Morrison, but I was almost getting used to it and we were definitely making progress.

The next job we tackled was Peggy's sills, which had experienced a chemical transformation from ordinary ferrous metal to something resembling Farley's Rusks. I ordered the new sills, which arrived at my house while I was at work. The quick thinking postie delivered them care of the Jolly Woodman pub, which was situated conveniently across the road from my house.

Although I lived close enough to almost touch the pub sign from my bedroom window, after work I took no chances and ventured into the Jolly Woodman to collect the sills. Of course, I could do nothing but politely stay for a swift half. After the seventh 'swift half', the conversation turned to my brown-paper-and-cardboard-wrapped package. Was it a fishing rod, a snooker cue or even a sniper rifle? I egged them on to further guesses. Curtain rails? A pair of antique hunting spears?

When I revealed that it was a pair of sills for an Austin 1100 Estate, the disappointment and puzzlement were audible. It was followed by, yes, you've guessed it, comments such as 'Oh my Dad had one of them. It fell apart after a year' and 'My brother worked on them in the factory – he only put half the bolts in on a Friday before heading off to the pub'. I felt myself holding a rictus grin but also not holding onto a need to stay any longer. I left the pub to a chorus of good-natured jeering. I turned and waved the sills at the cynical clientele as I clanged out of the door shouting, 'You'll see ... You'll see!'

Saturday dawned on the hunched figures of myself and Barry sitting on two spare tyres on a driveway next to Peggy. We were dressed appropriately, but perhaps not elegantly, for rolling around in the filth and grime under an old car. A young lady passing by stopped and stared. She looked firstly at the condition of Peggy and then her gaze fluctuated between myself and Barry. She rummaged in her handbag and it was only at the last minute I prevented her from making a monetary donation. We both

A monetary donation.

smiled at her and she shook her head, turned quickly and hurried away.

So, back to the sills. 'Where do we make the join?' I volunteered as a clarion call to action. Barry stirred briefly and grunted. Barry was able to make a simple grunt speak volumes. I deciphered that he was all for ripping off the originals at the return edge at the top where the door closes. He also raised the issue of blending in at the rear on a two-door vehicle. You might think that was quite a lot for one grunt, but you don't know Barry.

Anyway, someone had to decide where to make the first incision. Here the eloquence of Barry's grunting remained silent. He looked away nonplussed. I looked at Barry looking away, looking nonplussed. I turned the problem around in my mind. Joining the sill at the pressed top would leave me with a joint that would be hard to finish to the original factory standard and it would be in a very visible place. On the other hand, it had to join somewhere. The top three-quarters of the sills on both sides were absolutely sound so I figured I could get away without removing the top of the original sill.

The solution came to me in a flash! If it was done well there was no reason that the sills could not be joined at the start of the lower radius curve on the tubular lower section. If the panels and the receiving material were cut accurately, the weld could be lost in the radius curve slightly under the line of vision. This would leave the factory finish on the top of the sill intact with no loss of strength.

I explained this to Barry and the few muscles he had ever used in facial expressions failed to move. Sighing, I started to explain again and instantly felt like Stan Laurel having a good idea but being asked to 'Tell me that again'

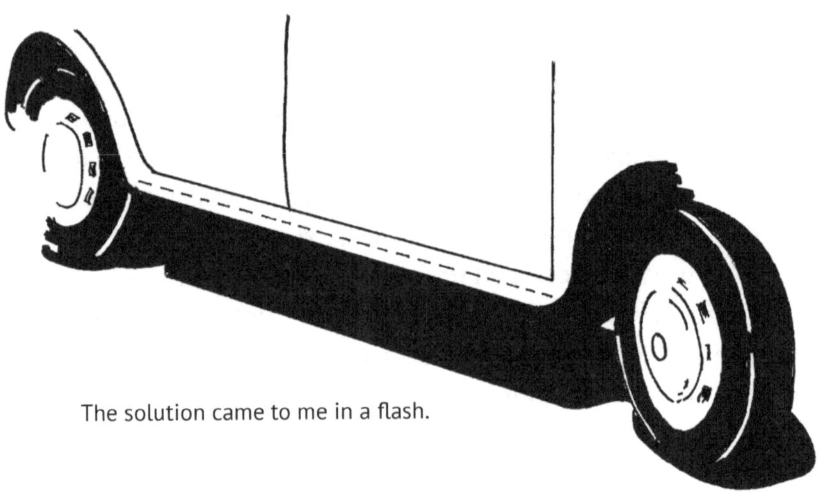

The solution came to me in a flash.

by Ollie. I explained again in true Stan style by completely mixing and mangling my words. There was a brief silence until Barry said, 'That's a good idea.' I almost found myself answering, 'It certainly is!' but I decided he probably wouldn't have got it.

Both sills were added in the fashion I have explained. We checked the strengthening membrane ribs inside before final closure and they were in remarkably good condition. We painted them, replaced the inner sills to the floor panels and sealed them. We stood back and looked at our work. We had succeeded! Barry looked at me, smiled, farted and then drove off.

Just as Barry's vehicle left the driveway, a small, balding head appeared over the dividing hedgerow. It was a seemingly friendly chap from the bowling club next door. He invited me over for a beer at their next match on Sunday. Now, I have to admit that my interest in bowls is such that if I were Francis Drake, the defeat of the Armada would have happened two hours earlier. However, it seemed a nice gesture, so I accepted.

I debated long and hard with my partner Marilyn about what would be suitable attire for a bowls club on a Sunday afternoon. After many unfeasible suggestions, I said that perhaps I should just be myself. Marilyn looked up and shook her head. She was right. Being myself would be a disaster.

I duly arrived at the bowls club prepared to be respectful and polite but to stay for as little time as possible. In my mind, this was a diplomatic mission as the bowls club was right next to where I was restoring Peggy. I needed them to be on board with the project, or at least indulgently tolerant. I had left the car behind so I could have a beer. This cunning strategy I found had always helped to disguise my terminal boredom. It was all going swimmingly by three o'clock and I had managed to smile warmly and offend no one.

All at once, someone nudged me to be quiet as the last end was on and it was a tight game. The last ball was being bowled. I watched the bowl gracefully winding its way slowly across the smooth green lawn. It was about to nestle right next to the jack in regal and righteous victory. I looked around: all were transfixed by the sporting drama unfolding before them. To gain the full benefit of this moment, I then actually turned round to face the action. As I did so, my foot brushed lightly against the rack that held all the spare bowling balls. I watched in growing horror as the rack slowly tipped over, releasing all of the bowling balls at a majestic but determined pace. I counted four of them rolling across the brick patio that fronted the green. My horror reached new heights as the lead one bumped over the edge and plopped onto the green. It seemed to defy physics, gathering pace to not only successfully intercept the 'winning' bowl but then go on to neatly nudge the jack so it nestled right next to the away side's bowl.

There was a stunned silence. It was the deafening sound of applause not happening. I felt a sudden urge to fill the silence. Losing control of my hands, I nervously applauded. I quickly realised that everyone had turned to stare at me. I shrugged my shoulders and adopted the smile of a naughty 4-year-old child who has been caught colouring in the cat with Dulux Weathershield. Seeing that my smile had not been well received, I raised my drink to my lips. Perhaps I could hide behind my pint of bitter. Before I could take a much-needed swig, I was seized by the elbow and bustled over to a corner by the home team captain.

It has been my experience that sentences that begin with 'Look here you …!' are probably not going to end with words of rapturous praise, and so it was. Graham, the club captain, pulled me away from the growing noise of heated arguments on the patio.

I could almost see the steam coming from his ears. With a red face and a furrowed brow, he informed me in a furious whisper that the reason I had been 'invited' was that the committee wanted to have a stern talk with me. They believed in being tolerant and neighbourly, but they weren't happy with me mucking around with that noisy and dirty old heap right next to their bowling green. They had spoken with Dennis who owned the driveway where Peggy was parked. I took my leave of the bowls club muttering, 'I've been thrown out of better clubs than this, you know …' I had a lot to think about. So, of course, I headed for the pub. As I sat down with my first pint, a text arrived from Dennis, 'Can you move it, mate?'

We can all remember hearing the well-worn phrase 'Cheer up, things could be worse'. Bearing this in mind, I forced myself to cheer up and, sure enough, things did get worse. The next week at work, my team were disbanded due to a withdrawal of central government funding. A month later, I was jobless and Peggy was homeless.

That Friday, I sat in the pub with my partner Marilyn and a few of our friends. The conversation was wild and wacky as usual, but for a few moments, I was lost in thought about my situation with Peggy. Unaware, I blurted my thoughts out loud. 'There is nothing for it, she'll have to go!'

My friend sitting opposite nodded wisely and chimed in with, 'I knew you'd make the right decision. Marilyn's a great girl but where are you ever going to get a car like that again?'

That settled it. Peggy was going to stay (and Marilyn too, of course). I would just have to find a way.

8

# A Game of Vehicular Russian Roulette

Many and varied are the slings and arrows of outrageous fortune that imperil anyone attempting to restore or keep their classic car on the road. They don't just come in the form of a rounded-off nut, receipt of a 'wrong handed' part or 2 inches of murky water in the footwell. Incidentally, the last one actually did happen to Peggy during the torrential rainstorms that covered East Kent before Christmas!

The privilege of owning a classic car does not prevent all of the personal disasters that happen in our lives. Personal difficulties such as ill health or the loss of family members can stop us in our tracks. In September 2011, I unexpectedly lost my job. This made it difficult to continue providing for my family and to push forward with my work on Peggy. I did consider selling Peggy for spares or repair, but I realised that this would be a knee-jerk reaction that probably wouldn't help my finances too much in the long term. So, I decided to back myself to get out of this with Peggy, Marilyn, my mortgage and my bar bill intact.

I applied for many jobs. Indeed, I am pretty sure I was shortlisted to be Queen at one point. However, despite countless interviews, I still hadn't landed a position. Frustration was creeping in.

In the meantime, you may remember, I also had to move Peggy from the driveway next to the bowls club. There was no way I could afford a garage, but Marilyn helpfully chimed in with information about a relative of hers who lived not too far away and had a driveway with no car. She also mentioned that he could do with a few quid to help him get by. So, off I trotted and spoke to Phil, who was, and still is, a lovely guy. The deal was done and all that was required now was to move Peggy from Bromley to West Wickham. It was a distance of probably no more than 4 miles but given Peggy's condition, it could have been 1969 and Apollo 11's mission to the Moon. As it turned out, the trip more closely resembled Apollo 13 but with a slightly smaller chance of survival.

A Saturday morning was duly appropriated for the move. The previous day, I had persuaded Barry the welder to at least put some metal into the spaces where the floor met the inner wheel arch. We also undid the front subframe lower mountings where they are fixed to the floor and the front bulkhead. I had some industrial one-eighth thick, 4×4cm square metal plates with pre-drilled holes that we placed on the inside of the lower subframe mounting holes near the floor. I needed to make sure that when Barry set off with his van and me in tow, the whole of Peggy would actually follow him. But it was all going to be fine. Barry reassured me that he had a special towing bar that he could bolt to Peggy and everything would be just great. No need to worry!

The fateful Saturday came and I arrived at the driveway ready to go. Barry arrived only an hour late with stories of escaped wildebeest blocking his route from Croydon. Well, that is what I thought I heard. In reality, I think he was confused by the appearance of a 'Stop'/'Go' man at some local roadworks. I was about to ask him whether the GO board had contained advice on an alternative route that took in Azerbaijan but I wisely decided to let the matter

An alternative route that took in Azerbaijan.

drop. After smoking several fags, and a rather detailed description of his cousin Lauren, who was 'top totty', Barry finally got round to bolting the tow arm to Peggy.

I waited for him to add the extra linkage, but he didn't. He just walked casually to the front of his van and started it up. I looked at the linkage. There wasn't a lot of it. The two vehicles were so close together that they looked like they were mating. Shaking that unwanted image from my mind, I hurried to the front of the van and told Barry that I thought the linkage left the vehicles too close together. He looked at me with the puzzled expression of someone who has had the origin of the universe explained to them over the course of six hours by Stanley Unwin. After some time he shrugged, unknitted his brows, pulled the door closed on his van, put on the hazards, lit up another fag and got ready to go.

I was a little concerned that at this stage Peggy had no braking system; only the hand brake worked. Hey, I told myself, people had travelled to the Moon without brakes.

We waddled our way off the driveway and onto the main road out of Bromley. The van gathered pace and so did Peggy. It dawned on me that all I was ever going to see on this journey was the rear end of Barry's van, in exquisite detail. Although I had put a battery on Peggy so the indicators could be used, this seemed rather pointless when all I could see was the hazard lights of Barry's vehicle. I had no idea which route he was taking to Peggy's new home. Consequently, every junction became a game of vehicular Russian roulette. I found myself reciting some scraps of Buddhist mantras I heard in a documentary and hoping that Barry did not encounter the same 'Stop'/'Go' man with the directions to Azerbaijan.

The last 2 miles felt like re-entering Earth's atmosphere. Barry had sped up to 40mph and I could not help imagining what the locals at the Jolly Woodman would say when they saw the 'And Finally' piece on *News at Ten* explaining what had happened to me. There is a wonderful piece in *1984* where George Orwell describes the feelings of Winston Smith as the cage with the rat is put around his head. The idea being that eventually, the rat would become hungry and start to take bites out of Winston's head. Orwell describes how when we are faced with such horrible things, our brain switches off from reality to protect us. Unfortunately, Mr Orwell, I can tell you now that during the journey to Phil's driveway, my brain did no such thing.

Against the odds, we reached our destination, disengaged Peggy from Barry's van and slowly pushed her onto the driveway. I covered her up with a car cover, paid Barry and Phil and staggered off to the bus stop to make my way home. On the journey, I looked everywhere but at the road through the front of the bus windscreen. It had all been too much.

I landed up at the Jolly Woodman, where I regaled the locals with my retelling of the day's events. They embraced my tale with all the enthusiasm of someone forced to watch daytime TV. I didn't care – I was home and I had a pint in my hand. Now it was time to think. How do I fund this? I had some savings and a good record with my local bank. They were always advertising about how they liked to help people. I am sure that they would have seldom come across anyone who needed more help than me at that particular time.

So it was the next day that I phoned the bank and asked about a loan. The lady went through all the usual checks and I told her exactly what it was for. As she had no idea what the car was, and no interest in finding out, I spotted my chance and went into great detail about Peggy. I think I even mentioned Barry, which may have done the trick. She glazed over (at least I think she did, remember we were on the phone). In a last desperate bid to cling on to her sanity, she granted the loan.

Now, the more financially astute among you will have noted that at this stage I didn't have a job and that, yes, I was going to have to pay back the first instalments of the loan out of the money I had borrowed. Well, I thought, if it was good enough for the Chancellor, it was certainly good enough for me.

The next few weeks saw great progress on Peggy and, although Barry never actually turned up on time, he did manage to put in a new clutch. (I had previously tried to move the clutch actuator arm with no success.) Barry had the whole thing working in forty minutes!

A couple of days later, we were scratching about at the rust that was inside the engine bay. There was quite a ragged seam where the return edge creates a shelf just behind the carburettor 'dimple' and above which the brake and clutch cylinders sit. You may be familiar with this particular area and in particular the site just behind the carburettor and air filter. I prodded the seam thoughtfully. At first, I was contemplating recreating it with just a welded seam. I probed a little further and the extent of the problem became all too apparent when I found I was suddenly able to see the heater box. There was nothing for it but to peel back the bulkhead and gaze at the disaster that was the air intake and the sad ruin of the heater box aperture.

Looking back, I am surprised we were brave, or foolish, enough to do it. We did have one sudden brainwave in making the bulkhead good and preserving the dimple shape behind the carb. As Barry was looking around for a spark of inspiration to help shape this part of the bodywork, I spotted a piece of leftover inner sill. I realised that if it was reversed, the radius would be pretty much the same as the dimple and the part of the bulkhead that meets the return edge. We finished it up and filled it and it really looked good. Barry smiled, scratched his backside and left. Although scratching his backside was a regular event for Barry, smiling certainly wasn't. I knew we'd had a good day. The pub beckoned.

9

# Another Attempt to Defy the Laws of Physics

I have often wondered at the bravery of some people. Most recently, the courage shown by Alexei Navalny. If you have not come across this guy, let me fill you in with the main points. Mr Navalny had been challenging Vladimir Putin on corruption for many years and had faced time in prison on many occasions. In Russia, this 'occupation' also comes with frequent beatings and many times ends in a career-, and indeed life-, ending death. In 2020, Mr Navalny was poisoned with a deadly nerve agent called Novichok and was rushed out of Russia to Germany, where his life hung in the balance for some weeks. After making a full recovery, he then elected to go back to Russia, was imprisoned on false charges and dispatched to a labour camp, where, tragically, he died.

I often look at some of the vehicles taken on by our membership in the 1100 Club and I ask the 'Navalny question'. Why are you doing this? Okay, so Vladimir Putin is not renowned for his interest in an innovative, small, front-wheel drive family car that first sprang from the Alec Issigonis drawing board onto a track at Longbridge in the early '60s. This is just as well because I have a feeling that if the subframe setting wasn't to his liking, there is a strong possibility that the Austin design office would have been relocated to Siberia.

Without the resolve of Alexei Navalny, I began to question the progress I was making with Peggy. The fact that I was now spending many weekend hours on Phil's driveway waiting patiently for Barry to turn up was starting to take its toll. Many times, the weather was against us and so too were the neighbours. There was almost a recordable displacement of air from the frequent twitching of curtains. Often, our progress was slower than Barry's mysterious journey times. Logically, he should have found it much easier to manage his time as Phil's driveway was a full 2 miles closer to his flat than our previous situation by the bowls club. Nevertheless, I continued to marvel at the distortion of the time/space continuum achieved by Barry's van.

Despite the distraction of Barry's groundbreaking experiments at the cutting edge of quantum physics, some milestones were reached. The valance/skirt just below the rear bumper area on Peggy looked like Godzilla had taken a bite out of it. And not just a nibble – it looked like that giant lizard was really hungry. I sourced a new, old stock rear valance panel from eBay. Barry turned up in his four-wheeled time machine, chopped out the old panel, made good the return edges to receive the valance and seamed it into position beautifully. The panel itself was, I believe, from a Vanden Plas saloon, so it came with one or two different pressings and attachments. I considered that the chances of coming across a MK1 panel were slightly less than Mr Navalny becoming President of Russia and saving a penalty in the FA Cup Final (on the same day!). So, I let the inaccuracies remain, nicely hidden by the helpful rear bumper.

However, since then, progress had stalled. Something must be done, I thought. I had landed a job driving a post office van, which didn't pay a great deal but at least when the job was done the day was my own. The phrase 'things were looking up' was far too optimistic to describe my situation, but if I could conjure up one analogy it would be from the oft-used *Titanic* disaster – I had at least found the cupboard where the distress flares were kept.

I chatted about the situation with my partner Marilyn one night in the local pub. Not surprisingly, she came up with a really good idea. I don't want to sound like I'm blaming her, but Marilyn has an annoying knack of speaking at exactly the time I'm not listening. Luckily, this time I *was* listening. We believed Barry owned a sizeable garage, so maybe I could rent some space there to store Peggy and he would then have no excuse for defying all known laws of time and space. Furthermore, Peggy would be snug and dry and all of Barry's tools were stored right there. Excitedly, we developed the idea. Despite assuring me to the contrary, I knew Barry was never officially 'at work'. He had often asked for money to get through and I knew things were difficult for him and his family. What if I could set him up in business with some of the money I had borrowed from the bank? Barry could then legitimately earn while completing work on Peggy as the payback. I could achieve my goal and help a family get back on its feet. A win-win, as the saying goes.

The next time I was standing on the driveway at Phil's, I explained our idea to Barry. I believe Barry tried to smile but it seemed the muscles in his face had forgotten how to set the process in motion. In truth, I don't think he had ever had much need for smiling.

So, it was then that we made plans to move Peggy to her comfy new home. As before, Barry attached the tow bar. It was at such a length that if I opened the quarter light on Peggy, I could reach out and draw rude pictures in the dust on the rear panel of Barry's van. We set off with a jerk. Correction: we set off with *two* jerks, one in the van and one following behind very closely in the 1100. The journey presented some extra challenges compared to the last time we moved Peggy. This time, we were venturing through Croydon town centre at 2:30 on a Saturday afternoon. Would Mr Navalny have contemplated this? I think not, even if Mr Putin's agents were chasing him.

By the time we arrived at Barry's garage in Norbury, I had slowly morphed into liquid form. Barry had taken the decision to go much faster than our previous trip, assured by the fact that Peggy was held together with much more metal than before. I had found myself, in sheer terror, desperately trying to complete hand signals from both sides of the car at the same time. The last time I had travelled so fast without being able to see where I was going was in the back of a speeding ambulance.

The following week (after some days of intense therapy at the Jolly Woodman) I spotted an advert on eBay for a set of sliding side windows for an 1100 Estate. They had all of the trim fittings and locking mechanisms and were being taken from a '69 MK2 that was being broken for spares in Gloucester. When I acquired Peggy, she came with one-piece side windows on both sides that were jammed into place. Even I thought that was ridiculous. I figured that if there was a MK2 being broken for spares, the chances were there were more potential items of interest for Peggy.

I arranged a day to head down to Gloucester and then contacted Barry. He told me that someone had responded to some leaflets and adverts I had put about and they wanted a quote for welding on an 1100 in central London. Things were looking up (again). I felt hopeful emotions welling up deep inside me. On reflection, this was more than likely to have been the after-effects of the fifth pint from the night before.

So, the day was set fair. I picked Barry up in my Nissan Micra 1 Litre – a car from 1998 that was already 14 years old but absolutely indestructible. We took a look at the car that needed welding. Barry gave it a look of considered disdain. It was a look you might normally reserve for something hairy that had landed in your bath five minutes before you were heading out on a Friday night. Barry mumbled a few words to the owner and then scratched his backside. Soon enough, we left the scene and were on our way to Gloucester.

On reaching Gloucester, we began to make our way through the more densely populated areas. I noted with surprise that this appeared to be the place where many bizarre and interesting classic vehicles had come to die. Almost every house displayed a vehicle in some form of disrepair embedded in the driveway. Many of the cars I saw would have come under the 'interesting' bracket for most of us. Some looked like they had been victims of failed resurrections; splattered with crusty red oxide, through which new rust had begun to germinate. Curiously, the only house that didn't have one of these offerings on the driveway was the one where I had made the appointment to view the donor estate car.

I pulled up outside and rechecked the address. No rust-covered offering to the god of oxidisation was visible anywhere. Granted, there was a driveway, but it was only a slim entry at the side of the property. I knocked on the door and we were greeted by the lady of the house. I realise now that the scene she looked out on must have given her the idea that she had just received a surprise visit from Stan Laurel and Oliver Hardy. Barry was a rather large and baby-faced individual and I was not. Eventually, after an almost audible roll of the eyeballs, the lady turned and shouted, 'Pete!'

A rumbling voice that was at least 6 foot 5 answered from the backyard, and an impressively large, overall-clad figure beckoned to us with a giant, hairy paw of a hand. We stepped cautiously through the side entry. He took us around a neat and manicured lawn to a small path that led further into the garden. At the end of this path, a strange sight met my eyes. It took a few moments to realise what I was looking at. It appeared that a whole Austin 1100 Estate had been carefully taken apart and carried neatly through the 2-foot-wide entry, down the crazy paving between the prize-winning roses and placed in long, slim pieces at the end of the garden. As my eyes adjusted to the view, I could

'Pete!'

see there really was a whole car sitting there in the garden. How were we going to get this into the back of the Micra?

In the end, I took two completely refurbished subframes, two doors, a tailgate, a petrol tank, four displacers, a radiator and, yes, the side windows. Pete was a lovely chap and charged me much less than the parts were worth. The car he had scrapped was actually in better condition than Peggy, but Peggy had not been 'adapted' to fit through a 2-foot-wide entry. I quickly let that thought go.

Pete's wife watched from the window as 'Laurel and Hardy' tried to make all the parts fit into the back of the Micra. (For Nissan Micra, by the way, just think Model T.) This seemed like another attempt to defy the laws of physics. I informed Barry that we had to be careful, as the car was not strictly mine and actually belonged to Marilyn. Barry replied that this was okay because if the car wasn't mine, what did it matter? Barry's book of 'Logical Truth and its Practical Application' was probably not going to be a Christmas stocking filler that year, I felt. We eventually crammed everything into the Micra and gingerly made our way back. One hill on the way out of Gloucester nearly defeated the brave little motor, but later that evening we arrived in one piece at Barry's garage and deposited what I had come to think of as 'The Gloucester Marbles'.

No sooner was the garage door shut than Barry heaved himself into his van and sped off without a backward glance. It was late and my drive back to Beckenham from Croydon was somewhat easier. As the lights flashed by through the growing gloom, I saw people of all kinds happily going about their business and living their busy lives. I considered my own life and wondered what was so empty and unfulfilled about it that I had willingly spent my day in Gloucester scrambling around rusty, oily scrap metal. I thought of Peggy, and then I thought of Barry. I felt my foot pressing down on the accelerator – was the pub still open?

Flattening out a rather rusty roof.

Replacing a door panel. Only two more to go!

Nearly finished? Not really, but Peggy does have a new floor and bulkhead.

A spot of welding in the boot.

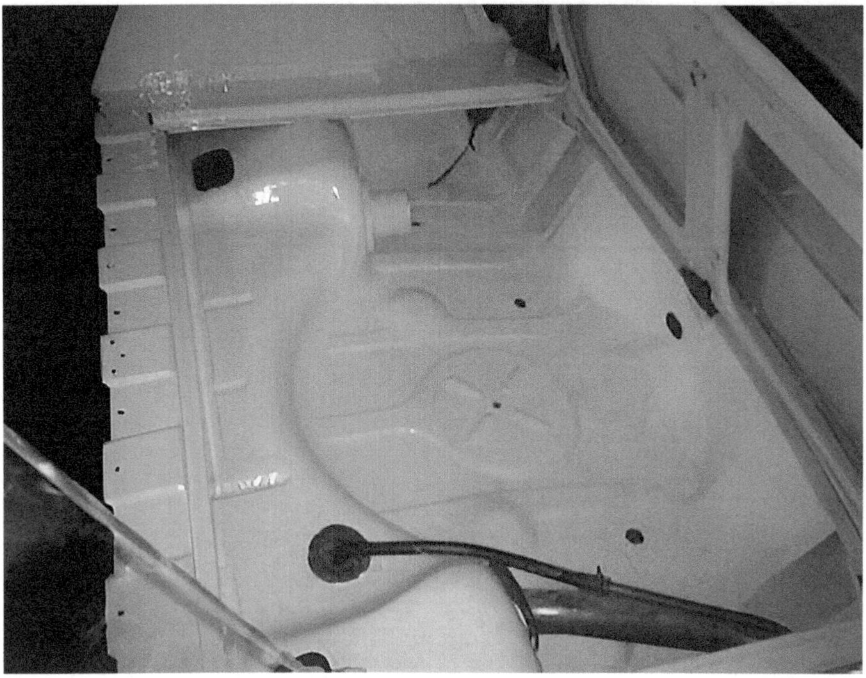

The boot is finished and looks lovely!

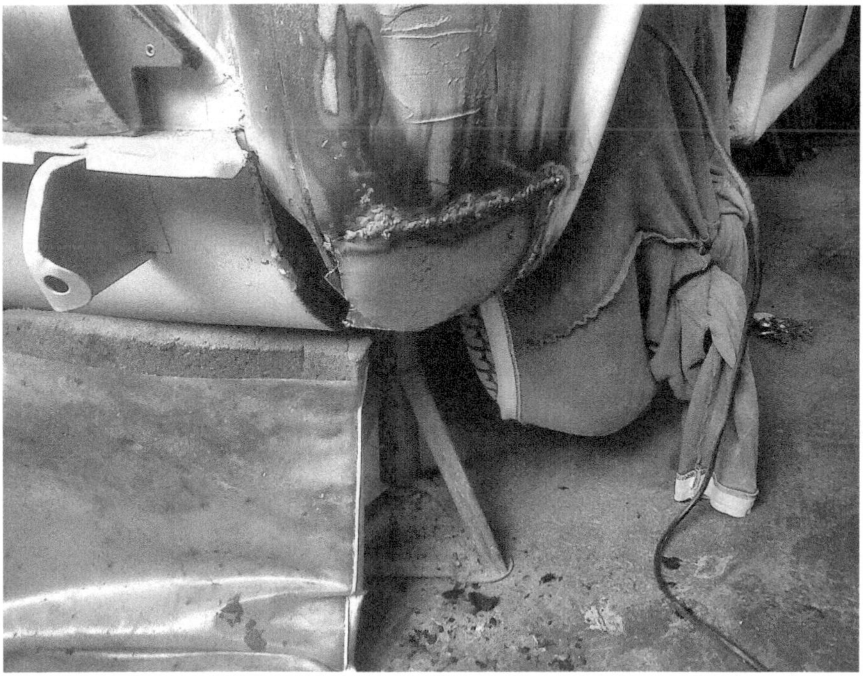

A new valance for Peggy.

Peggy's temporary home on Phil's front drive.

A shiny new floor!

A work in progress. I suppose we should get some wheels.

A nice coat of Snowberry white and putting the *Sun* newspaper to good use.

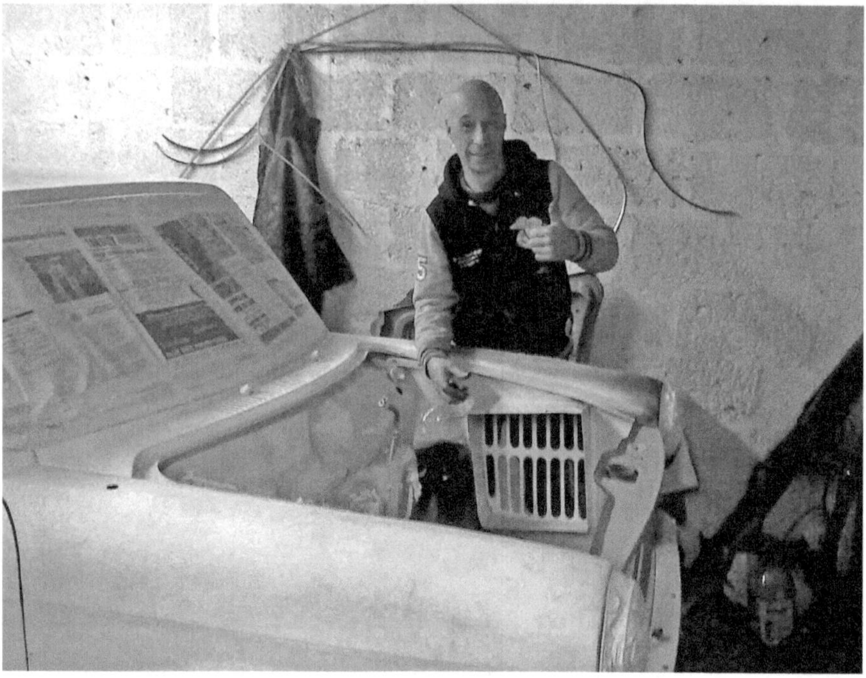

Something missing? Ah yes, the engine!

The engine, minus the head.

The engine in situ, with the head attached.

Peggy in repose in our home garage.

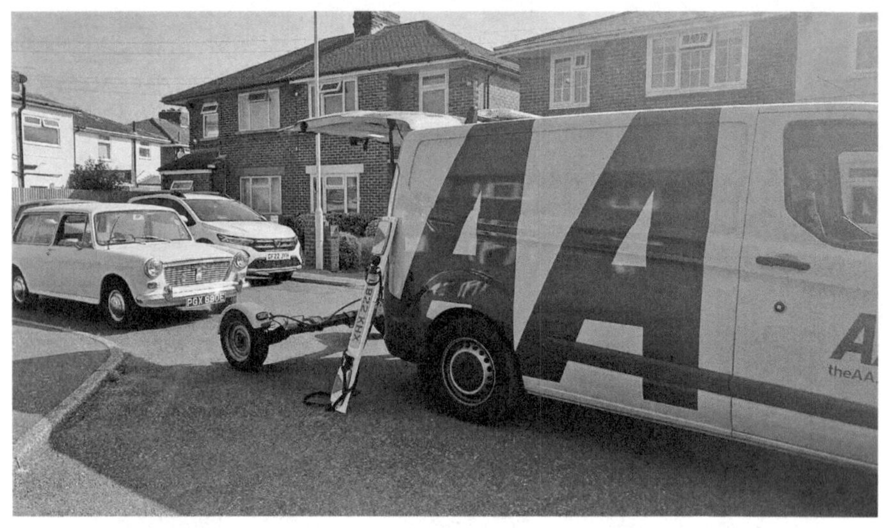

Peggy has let us all down. The man from the AA was far too young to know how to fix her fuel pump.

Lost on the way to Fredville Park, near Nonington, Kent.

10

# The Lights Go On, and So Does the Kettle

I am often reminded of the words of certain prophets or philosophers, either through my own recollections, from my partner Marilyn, or as often as not, as memes on Facebook pages. While intriguing, they never seem to me to be awfully practical. I wonder at times about what Confucius might have to say about removing the petrol tank on an 1100. What wise thoughts might the Pope have, whilst skinning his knuckles trying to remove a rear displacer in a hail of un-Popelike blasphemy? Or perhaps the Dalai Lama might have some thoughts advising against mixing cross plies with radials? I can tell you that there were many times that I reached desperately back through history for words to help me cope with the restoration of Peggy and, indeed, with Barry.

Life at Barry's garage presented new opportunities to move things forward, coupled with new opportunities to move things backwards. On the plus side, there was cover, and for the most part, protection from the elements. The roof to the garage had long since been enveloped in large tree branches, which made it look like it belonged to one of those abandoned and mysterious villages that nature had slowly reclaimed over hundreds of years. This meant that the only time the roof leaked was when it rained. On the downside, there was no mains electricity and power was supplied by a generator.

I soon learned that there was a distinct correlation between my arrival at Barry's garage and how low the generator was on petrol. Our routine went something like this: I turn up at the garage; no Barry. I phone Barry; Barry says he is on his way. Barry arrives and then tells me that his partner doesn't appreciate him and that his cousin is still a 'bit of alright'. This discussion usually happens over a cup of tea when Barry has cranked up the generator and lasts for at least an hour. Now fully informed about Barry's relationship and family problems, I skilfully move the conversation around to what we might get done with what little time remained of the afternoon. We agree on what we are tackling and Barry then plugs in his welder and angle grinder into the generator. No sooner do we start on the car than the lights go out, the generator coughs loudly and stops dead. With a sigh, I take a petrol can to my car and drive to the nearest petrol station, which is a mile away, and then make my way back. I pour the petrol into the generator. Barry then tries to crank up the generator but it seems to have acquired learnt behaviour from its master and refuses to start up. Eventually, a miracle; the generator vibrates into life. The lights go on, and so does the kettle.

The lights go on, and so does the kettle.

As you will have noted, this wasn't exactly thrusting progress but, still, things did get done. The engine and box came out and we removed the front subframe. Barry had somehow constructed a frighteningly unstable hoist to lift out the engine. At first sight, it looked as if it would be structurally compromised if it attempted to lift Marilyn's knitting box onto our kitchen table but, another miracle! It worked … just. This allowed us to clean up the engine bay and weld the seam where the inner wings join the main bulkhead at the cabin end. This is a problem area that we would not have seen or even thought of had the subframe not come out. This space is dangerously hidden behind the subframe towers and the area of rot was apparent all down the seam to the bottom front corners of the footwell.

The engine head was sent off to a local engineer for conversion to unleaded petrol. The front displacers were examined, cleaned and set up with a new roller foot kit purchased from Longbridge Spares, along with all new subframe and engine mounts.

This was all good progress, if sometimes a little painful. It often felt that I was acting as Barry's therapist while paying for the sessions myself. Barry had made me realise that it had been a long time since I had met anyone who I would call happy and well adjusted.

Before we put in the new front subframe in all its shiny black glory, I had to decide what colour Peggy was going to be so that we could spray the engine bay. My artistic sensibilities were torn. Peggy was originally a mid-blue, which was common for the early estates. The car that I had originally owned in the '70s was Snowberry white. Well, OOG 300G had been the inspiration for this whole endeavour, so nostalgia won the day. Snowberry white it was!

As progress continued, so, unfortunately, did the bills. Parts were not cheap and Barry had now decided to up his rate of pay. He did this not by having a frank and heartfelt, man-to-man conversation with me but by the more subtle method of doing less work over a longer period of time. I was made aware of this move by the length of time he took to let in new metal to the bottom of the driver's door, compared to the length of time it took to let in the metal for the passenger door. One took substantially longer than the other, even though they were identical repairs. One took one week and the other took four weeks.

I turned up one day at Barry's garage after he had asked me to come and see the heroic efforts he had put in on Peggy. 'I have been all over the car, grinding down the areas that needed work!' he boasted.

To be fair, this was partly true ... very partly true. If you have ever read any Sherlock Holmes books, you will know that the great detective can spot many indicators of recent behaviour of which the observed person themselves are not aware. I did not retrieve my magnifying glass, nor did I feel the need to don my deerstalker, but I did find myself channelling the famous sleuth. When looking at the floor and the dust that had been created, I noted that there were two clear spaces where Barry's feet had been placed. There were no corresponding clear spaces anywhere else around the car. There were two distinct possibilities. Either Barry had accessed some ancient wisdom and was practising levitation, or he had just sprayed the dust on the floor around the car. Using rational deduction and, of course, a previous knowledge of Barry, I deduced it was more likely to be the latter. 'When you have eliminated the impossible, whatever remains, however improbable, must be the truth.' Mr Holmes, I'm sure, must have had Barry in mind when he wrote this.

Dealing with Barry was getting harder, but what could I do? I confided in Marilyn, who could be relied upon to come up with a way forward, even if it was not the easiest route. Marilyn suggested that we invite Barry and his partner Kelly round to ours for a meal and there, in a more relaxed setting, we could talk things through and come up with a more rigorous plan to finish Peggy's restoration before it finished me.

I sat in the pub the next day and images of the yet to be arranged evening were flashing before me. What had we done? In what remained of my functioning mind, I imagined a chaotic scene of shrieks, rude gestures and food flying everywhere. It would be like a hellish version of those old adverts for PG Tips, and our voices would be dubbed by Johnny Morris. *Animal Magic* indeed.

11

# Where Was Charlton Heston When I Needed him?

'And so it came to pass ...'

Words I often heard from the great biblical epics shown on TV during the holidays. As I recall, these 'epics' often featured famous actors such as Charlton Heston, James Mason and Elizabeth Taylor. We mere mortals stood (or sat on the sofa) in awe of these revered thespians and the powerful figures they portrayed.

The arrival at our house of Barry and his partner Kelly never quite reached those heights. In truth, I had no great expectation that it would. Despite this, I had already looked out of our front window for a 'sign' or portent of some kind; perhaps a burning bush or something of that nature. Indeed, even parting our local river, the Beck, would have been difficult to record, as you can walk over it in some spots without any great difficulty. However, there was a sign! It was not biblical, but financial in nature.

Kelly knocked on our door at the agreed time. I opened it up to greet her and noticed Barry lurking beyond her shoulder. He had just parked and climbed out of a brand new and very shiny Nissan NV 400. Now, I realise that the sentiments expressed in the next few sentences may reflect more on me than on Barry. However, all I ask is that you consider the fact that I was probably Barry's sole source of income and, at times, support. While I otherwise would have been ecstatic at Barry and Kelly's evident progress in acquiring a brand new top-of-the-range vehicle that I could not afford, I was somewhat troubled by the notion that I had probably, single-handedly and

A band new and very shiny Nissan NV400.

unwittingly, funded that progress while at times being unemployed myself.

We all settled down at the table and exchanged pleasant niceties. The evening actually went rather well, apart from a nagging feeling I had that the main topic I wished to discuss was being expertly dodged. I needed to talk about our progress with Peggy. I needed to know important information such as time spans and costs. Barry and Kelly just needed to know what was for dessert. After an evening of unsuccessfully trying to manoeuvre the conversation towards Peggy-related topics, I eventually gave up and we said goodbye to Barry and Kelly, letting out a sigh of relief as they left.

The plain truth was that I was no further on with the Peggy project. I considered pulling the plug on my relationship with Barry, but how could I? He had me in checkmate. Peggy was still not reconstituted and was sitting on axle stands in his garage with all her accompanying bits. What to do next? My first impulse would have received hearty approval from all of the main characters in *Goodfellas*. I talked it over with Marilyn and she advised a more subtle approach.

Of course, I disappointed the fictional gangsters and listened to my wise and sensible partner. I took the approach of encouraging Barry by way of motivating him with praise. Praising him for being only half an hour late at the garage was difficult, but generally what he did do was actually good.

We lowered the engine back into Peggy on a tired-looking rope and Barry's hoist, which appeared to be made of discarded Meccano pieces. I hope the readers will forgive me when I boast that we managed this feat without scratching the newly sprayed engine bay. It was not an easy ten minutes, as you will all know. The engine, dangling on the end of this very frayed rope, reminded me of the 'grab' machine at the amusement arcade that allows you to pick up your prize with deceptive ease. As soon as you manoeuvred it into position to be deposited into the collection chute, when you are absolutely sure you have beaten the machine, when you are beginning to formulate your look of smug superiority over everyone else in the arcade, the prize drops with a sad thunk. It not only drops, but your 'prize' proceeds to bury itself in the heap of plastic detritus at the bottom of the glass container. I was prepared for the engine to crash with its own thunk into the bulkhead or inner wings at any moment. It didn't.

I realised that I had now become ecstatically happy that there *hadn't* been a disaster and this was now the level of expectations I had in my life, thanks to Barry and Peggy. Where was Charlton Heston when I needed him? Probably winning all the prizes at the amusement arcade.

Things were still not easy but we did make progress. With the engine now in place, new mounts and all other items were checked off. There was a new radiator, a dynamo, a change in polarity, and a new old stock well for the clutch and brakes (okay, they were plastic, but practicality won the day on that one). A new, old stock single carb was added. The dashboard slotted in, albeit without a properly working strip speedo; something that I dealt with later.

It was now 2013 and finally (in the words of Oliver Hardy) 'we were getting someplace'. Having accomplished a great deal at the front end of the vehicle, our attention now logically went to the rear. With my reliance on Barry almost over, I needed to consider whether dropping the rear subframe out was really necessary. It was, but I still considered the situation carefully. If Peggy had been sitting in a field for six years, it occurred to me that it may be very unwise to start the car up and suddenly pull all that accumulated sludge through the fuel system and our newly fitted carburettor. So, I suggested that rather than take the subframe out, we could take out the petrol tank, clean it up, and while it was out, we could inspect the subframe without needing to remove it.

I congratulated myself on such logical thinking. I was not only finding a practical way to deal with a restoration problem but also a way to sidestep any prolonged reliance on Barry. Many of you will be familiar with the Herculean trials of refitting a petrol tank on one of our cars. During this endeavour, I became cognisant of the exploits of the plastic surgeon Archibald McIndoe and his 'Guinea Pig Club' during the Second World War. By the time Barry and I had cleaned the petrol tank and finally tightened home the last bolt on the tank, the garage floor was littered with bits of skin and splashes of blood. I am fairly confident that even the renowned Dr McIndoe would have struggled to return our hands to any semblance of normalcy.

And so it came to pass (remember that phrase?) that, sitting in the pub later one evening, I considered that Barry and I had indeed mixed our very blood and genetic material in the service of Peggy. I wondered how many pints it would take to remove that thought from my mind. On the other (bloody and bruised) hand, it was quite epic and biblical in its own way.

12

# Looking Down the Wrong End of the Telescope

Many of you will be familiar with the managerial 'merry-go-round' of Premiership football; the cut-throat environment where every manager and player is under constant pressure. You will hear phrases such as 'results based', 'stats driven', 'meeting or surpassing expectations', and so on.

Well, despite being in the similar pressure cooker environment of classic car restoration, Barry was blissfully unaware of any of these things. Even if he were to become aware, I don't think he would have been able to understand them, or indeed spell them. If I were to compare myself to a Premiership manager, I would have to say that I had lost the dressing room, and I was unlikely to find it again. This is despite only having one player to manage. It is true to say as formation options went, I was rather limited. Man management tactics, like giving praise even where there was little or nothing to praise, had failed consistently. In fact, they had just produced more non-praiseworthy outcomes. However, as I was, by default, also the board of directors, I gave myself another chance to put things right.

Progress on Peggy had ground to a halt. I had gone to Barry's garage on three occasions, armed with all the supplies I needed to clean up the original front displacers. I was also in possession of a new front roller foot kit for both sides, which I had purchased from Longbridge Spares. On all of those three occasions, Barry had suddenly cancelled. Each one was another Sunday wiped out and further slippage in the progress of Project Peggy.

Then one day, Barry rolled up in his bright, shiny and expensive Nissan van. I felt like Ben Gunn being visited by Long John Silver; I was glad to see him but a little apprehensive at the outcome of his visit. Barry didn't make any excuses for his previous misdemeanours but instead filled me in on the cousin that he fancied and his wife's latest exploits at a Croydon nightclub.

While doing this, Barry made a move to start the generator to make some tea. In attempting to start the generator, he found he had no petrol. I did not bother to mention that I had already discovered he had no milk, or, indeed, any tea.

With a sigh, I left to do my usual run to the petrol station. As I drove away, I considered my position. I decided the football metaphor was still a useful one. It was difficult, but motivation was the key to getting the dressing room back. In a spark of inspiration, I reflected on the many conversations I'd had with Barry. It might have been a stretch to call them conversations, as that implies some kind of two-way process. My contribution was minimal and on many occasions, completely unnecessary. These 'conversations' often centred around his wife Kelly's inability to understand him. In truth, I sometimes found it difficult myself. Kelly clearly had never appreciated Barry's love of vehicles, and classics in particular. Even on the days Kelly had been at Barry's garage she had never lusted over his three-quarters completed MK2 Mini half as much as Barry had lusted over his cousin.

In a moment, I had it. I had been looking down the wrong end of the telescope. Rather than internally berating Barry's disturbing fixation with his 'fit' cousin, I should use this situation as a positive. (See what I'm doing here? Putting a positive spin on a negative situation is a classic Premier League management tactic.)

On my return from the petrol station, we cranked the generator into life and Barry finally made a cup of tea. I didn't complain about his lack of activity regarding Peggy but instead led him into a conversation about Kelly and his attractive cousin. I threw in my observations about Kelly's indifference to his work on vehicles and casually asked Barry if his cousin liked old vehicles. He replied in the affirmative with a laugh that was eerily reminiscent of the late Sid James. I was onto something.

During our session reassembling the front suspension with new roller foot and bushes and cleaning up displacers, I suggested that Barry should invite his cousin to the garage to see how brilliant his work was. I also ventured that she would be mightily impressed by it. He agreed and our work began to move smoothly on. The displacers were back in situ and ready to go. With the use of a grease gun pump, we were able to pump up the suspension. All joints held firm with just a slight weep on the offside front displacer union. I resisted the temptation to give the joint one more turn and we duly let the car down, remaking the joint with plumber's boss white. This did the trick. Peggy was now standing on her own four feet. It had been a long time. We arranged to meet the next Sunday to attempt to refit the dash and heater box. The scene was nicely set for my Machiavellian plan to take shape.

I arrived that Sunday at the arranged time. I had taken the precaution of coming laden with petrol, milk and tea. With impeccable timing, Barry arrived half an hour late in his shiny van. This time, however, he was accompanied by the aforementioned cousin. I have to say now that Lauren was (and probably still is) an extremely attractive woman. Barry introduced us and quickly took Lauren by the hand, showing her the garage and walking her around his three-quarters finished Mini. Lauren seemed suitably impressed. It was working already. As Barry was showing her the cones and front ball joints, she seemed to lose a little interest and turned round to look at Peggy. 'Wow, is this yours?' she asked me.

'Er, yup,' I replied.

'What is it?' Lauren asked.

I was tempted to glare in the direction of Barry and shout 'Unfinished!' but I resisted. As I explained to Lauren what Peggy was and how she had been acquired, I could still hear Barry's muffled voice coming from under the front wheel arch of the Mini. Eventually, his one-sided conversation tailed off and I suddenly became aware of him standing behind us as I talked to Lauren about Peggy. His

Wonderful, I thought, it's working!

jealousy engine immediately engaged gear and he acceler-ated quickly to possessive overdrive. He took Lauren by the arm and whisked her around Peggy, telling her what he was going to do, how he was going to do it and how fast he would be doing it. Wonderful, I thought, it's working!

I got to the pub later that evening and sat alone with a quiet, self-satisfied smirk on my face. I thought about what could possibly undo my plan. The arrival of Kelly at an inopportune moment? Not likely. I had turned things around. I had the dressing room back (albeit with a comely new trainer). It was time added on and I was a goal to the good. Even VAR could not defeat me now. Could it?

13

# Whodunit?

If you have ever read one of those 'whodunit' novels or watched a play or film about a murder, you will recall that you are encouraged to work out who the murderer is using clues sprinkled along the way. You will also recall how infuriating it is when the author inserts another character you had not bargained for right at the end of the tale. This happens just when you thought you had deduced that the illegitimate daughter of the Murdered Duchess had left her native Switzerland to get a job as a nanny at Slovenly Manor, and started an abusive relationship with the late Duchess's son who was, in turn, being blackmailed by the local postman (who had a record for petty theft but had been supported by the local vicar who used to be a pharmacist and had access to the poison that killed the Duchess). Stay with me, readers. When the stereotyped cast of suspects is gathered together in the ornate library at the end, in walks the bespectacled, mouselike secretary whom you barely remember uttering a single line in the previous 492 pages. It turns out she was the killer all along!

If you ever feel frustrated at such events, then you will completely understand the way I felt with the arrival at Barry's garage of his wife, Kelly. My carefully laid motivational honeytrap set for Barry via his attractive cousin Lauren was now in jeopardy of being undone. If you are shaking your head in puzzlement, go back and re-read the previous chapter. The writing doesn't get any better but it will help you understand this bit.

It was on a Sunday that I dragged my work and world-weary self up to Barry's garage. The task for the day was the tailgate. Parts for our cars are generally in pretty good supply. Two-door and four-door panels are always sourceable. I even managed to find a replacement MK1 strip speedo from one of our friends in Malta (it came in its original box!). However, how many tailgates for a traveller or estate do you see on eBay or classic motor sites? I did have an MK2 tailgate that I sourced from Gloucester but it was relatively scabby and unkempt. There wasn't much to choose between its metal content and that of the original.

After a chat with Barry, we decided to attempt to repair the original tailgate. The problematic areas of the tailgate were the bottom lip and the return edge, which were barely a ghost of the shape needed. The part of the pressing that receives the rear windscreen looked like King Kong had been chewing at it. Nevertheless, I had confidence in Barry and his ability to remake the areas needed. I had to; there was no other solution.

We sat in the garage with the kettle burbling merrily away. It occurred to me that if I had measured the energy use from the generator, that kettle would have spiked higher on any graph than the welder. A knock came at the garage door, interrupting Barry's illuminating top ten countdown of fit birds in the Croydon area. I was not altogether sorry this lecture had been terminated early as I was running out of faces I could pull that contained any semblance of sincere interest. The garage door opened and in strode Lauren, who most surely would have featured towards the pinnacle of Barry's 'top ten of Croydon'. Lauren smiled brightly and said 'hello', but before she could talk to me, Barry immediately sparked into action. He took her over to the tailgate and explained exactly what it was he was going to do. Lauren asked, 'How are you going to do that?'

'Slowly,' I mumbled inadvertently. Barry gave me a puzzled look and then carried on talking to Lauren, deciding that he must have misheard. I think it was when Barry started telling Lauren how to gauge the metal he was going to use and the type of welding rods needed that I noticed the expressions on Lauren's face coming to resemble the ones I had been forming during Barry's earlier lecture. With just our faces as a guide, only Lauren's own mother could have told the two of us apart.

I was heartily proud of the fact that I knew Barry would take no notice of Lauren's expressions as he carried on spitting out words like a runaway fax machine, spouting all manner of incomprehensible technical information. My boy was doing good.

After finishing with the grinder, he fired up the welder, and we were on our way. I felt like Peter Cushing as Baron

Frankenstein; with a bolt of blue light, my creation was being given life. Lauren stood back at a safe distance and chatted with me. I guess one thing that would not have been so popular in the clubs and pubs of Croydon would have been pockmarks on your face from flying shards of red hot metal. Lauren then told me that her husband was arriving any moment as he needed Barry to do a job for him. The echo chamber in my head reverberated with an inner shout of 'WHAT?'. This wasn't what I'd planned for.

On cue, Garry, Lauren's husband, entered the garage and started up a conversation with Barry. Barry downed tools and put the kettle on. I picked up the grinder and carried on. Soon enough, the garage door rumbled open again and I noticed that Barry's wife Kelly had entered the fray. In the time it took them all to drink tea and talk about the merits of certain local nightclubs, I had trimmed the weld in the screen area, sealed and primed it.

I then walked towards my disinterested audience and in a voice borne out of sheer frustration and no little anger, I shouted, 'We need to get the screen in!' I was satisfied to see that they all jumped to attention at this. I now realise it was not in response to my forthright encouragement to progress but more possibly in response to the fact that I had in my hand the still rapidly spinning angle grinder.

Barry and I worked to get the screen inserted. We eased nine-tenths of the screen rubber into place but that last inch or so just wouldn't go. Even with the twin powers of brute strength and ignorance, we did not succeed. Anyone listening in from outside the garage would have heard a confusing and alarming jumble of phrases: 'Use levers ... Let's try it another way ... Hold it from the inside ... Heat the metal … Don't heat the metal … Bend the metal … For f*ck's sake, *don't* bend the metal!'

Eventually, Lauren, Kelly, Garry, and all pitched in, and with wooden strips, pieces of string, grease and many hands, the rubber started to capitulate and slide reluctantly into place. I am fully convinced that in the end it was the last piece of foul and abusive language uttered by Lauren that eventually persuaded the retaining rubber that further resistance was futile. We all stood back and heaved a sigh of relief. Lauren and Garry left, and shortly afterwards, so did I, leaving Barry to explain to Kelly why he had invited Lauren to the garage.

'We need to get the screen in!'

I retired to the pub to contemplate the afternoon's events. The tailgate screen was in. However, I had noticed that a part of the rubber was slightly kinked, but I thought it wiser not to revisit the issue. To this day, ten years later, that piece of rubber doesn't look great, but on the other hand, it has never leaked. I will get around to replacing the rubber this year, I think.

So, I had wound up with more characters in my plot than I had anticipated. Whodunit? In the best traditions of Agatha Christie, I guess they all did.

14

# What Next?

Many of you will be familiar with the problems we face in restoring our vehicles. Handed parts, parts that change from MK1 to MK2 to MK3, left-handed threads, etc. The list isn't endless, but it is long. The one problem I know you will all have confronted starts off like this: 'Right, we need to replace this part!' Next, you remove it and look behind it. 'Oh ... right. Well, now that is off, I might as well have a look at the part underneath. It probably needs looking at and we will only have to get it all out again if the bearing goes ...'

Sounds familiar? I thought so. And so it was with Peggy. The original intention was just to get Peggy back on the road with the minimum of intervention. If you have read the previous verbiage, you will know that this was an optimistic stance bordering on mental illness. 'Minimum' was never a description of anything that happened to Peggy during the nearly four years it took to get her back on the road.

Despite this, by 2013 we were getting there, and in the end, almost everything was stripped off the car and either cleaned up, refurbished or replaced. The only thing that didn't come out was the rear subframe, and that was mainly because we took the petrol tank out and, by doing so, we were able to have a good look at the condition of the subframe. It was also because it was one of the last things we did and both myself and Barry had run out of the naive sense of adventure needed to take it on. I had also run out of tea, petrol and money.

One Sunday afternoon, after a long session putting the heater back in and jigging around with the dashboard and its padded top insert, I hauled myself back upright. You will all know that being upside down in the footwell for the best part of an hour is probably something that neither Houdini nor Yuri Gagarin would have undertaken without a few pints of 'Old and Filthy' inside them. Once the blood in my body had gone back to the right places, I considered that the only remaining body pieces that needed to be replaced were the two bottom rear corners of the car. I looked around for any replacement panels but I knew it was a forlorn hope. The rear quarter panels and wheel arches were good and inserting a whole rear panel would have been unnecessary butchery. So it came to pass that Barry fabricated the rounded ends and let them in. Even now, nearly ten years later, you would have to look closely to see the corners were not original.

Barry had done a good job and then spent the next Sunday reminding me just how good at welding he was.

I sat in the garage feeling like I should have created an award just for him, and perhaps had it presented by Lauren. He would have liked that. As Barry chuntered on, I imagined his garage as the stately Gielgud Theatre. I was on stage looking out at all the great and good of welding who had made their way in along a red carpet (full of burn marks, of course). The assembled worthies would be sitting expectantly, their other halves glamorously dressed in grease-covered boiler suits and goggles. I would come to the microphone (suspended by a cable tie), clear my throat and announce: 'Ladies and gentlemen, the award for most unpunctual welder goes to … (now reading in reverse order) YrraB- er … I mean – Barry!'

I imagined Barry taking the long way round to the stage, making a call from the upper circle informing me that he was nearly there but being held up by an ice cream vendor. What would the award look like? I know! A silver displacer mounted on a steaming kettle!

At the end of one less than frantic session of Peggy work, we stepped back and looked thoughtfully at the car. What next? We looked at each other and we knew the answer – paint! Barry uncovered a spray compressor that I had never suspected was in the garage. He then uncovered a spray gun that looked like it had last been used for making images of hands and bison on caves in France. In my opinion, it was not salvageable. It was only ever going to plop out sloppy globules of paint that might have been useful in a paint-balling battle. Peggy would have looked like the exterior of a Dalek, or a piece of art created by Yoko Ono using a large pot of porridge.

I had decided on Snowberry white as that was the colour of OOG 300G, my original estate that I had owned back in 1976. As Peggy was currently a dark blue, this met with a strange expression from Barry. As he was full of strange expressions that were usually a precursor to wind, the arrival of Kelly, or the need for money, I ignored him.

To affect Peggy's makeover, I needed to purchase a new spray gun, some primer, paint and thinners. This was not going to be cheap. I nervously peeked at my bank balance through half-closed fingers. It was anything but balanced. In fact, it was so unbalanced that I should have referred it for counselling.

Sometime earlier in the year, I had bumped into an old workmate I used to play in a band with. As often happens when drink and failed musicians meet, we promised to get together to do some music. Of course, as with many, many failed musicians, it never happens and the public is saved. Despite knowing this, I phoned my friend Ken and asked did he fancy doing some gigs. He hid his enthusiasm incredibly well. After a couple of rehearsals, which we appropriately renamed 'reversals', we got booked into some pubs. Despite spirited opposition from many of the customers, we managed to earn a fair few quid. I soon had enough for the paint, thinners, spray gun and Barry. Incidentally, I still play in a band with Ken and we have got quite a following, which is a far cry from the early days. There was one gig where, if the band had picked a fight with the audience, the band would have won by outnumbering them.

Back at the garage, the painting began and, as the primer spread over Peggy, she finally became the shape and style

I was helped home some hours later.

that Italian design had intended. I have often watched classic car shows on TV, only to switch off when the featured car is a Lamborghini Satsuma or a Jaguar XYZ or something equally exotic. The cars that always impressed me were the ones created by the guys who were given the design brief: 'Make it as small as you can, make it seat fifteen people and 3 tons of luggage. Oh, and it must do 35–40mpg and look like sex on wheels.'

That night, I sat in the pub and contemplated all the trials and tribulations of the Peggy journey so far. I was helped home some hours later.

15

# The Fatal Lure of the Heroic Failure

I used to do a lot of running. When I was younger, always being too close to the bus or train departure time meant that it was part of everyday life. As I grew older, I took up running as a hobby, a fitness kick and a de-stressing mind release. Every week, I would challenge myself to run a little further. As I got older, this became less important. The little voice in my head grew bolder over the years. You know, the one that whispers: 'Why are you doing this? Do you know how ridiculous you look? Turn back now before you get too far from home ...'

It often made me think about athletes that pushed their boundaries and wonder did they have to do the whole thing just to prove to themselves they could do it? Roger Bannister, on his last lap, could have just stopped and quite legitimately said, 'There you go, I told you I could do it. Just clock in the time I was on for, mate, and I'll see you in the pub afterwards.'

So it was with Peggy. After nearly five years we were almost there, and yet a strange feeling came over me. Did I have to carry on and complete this to prove to myself I was able to do it? Able to battle against the odds and all the slings and arrows of outrageous fortune and, of course, Barry. I was falling for the fatal lure of the heroic failure. Instead of people looking at me as a smug, self-satisfied individual who owned a nice shiny old car, I could tell my epic story at the Pig and Whistle of the gods being dead set against me. I would explain that I had given it my best shot while garnering sympathy and understanding from those intoxicated enough to still be listening to my story.

I was pulled from this philosophical malaise by the practical job of adding a top coat onto Peggy. Barry assured me he had sprayed cars before and, although I did have some spraying experience, it was mostly with weedkiller. Barry sourced a company that knew their stuff and we plumped for a two-pack finish, which Barry explained would be a harder and more durable result. As I previously mentioned, I had decided on Snowberry white. A word of warning for anyone going for this colour: many paint providers will supply you with their idea of Snowberry white. You should be prepared for a chalky cream finish that will instantly tell you that it is *not* Snowberry white. However, the company that provided us with the paint had it spot on. One night, while I was fighting the voices in my head that were telling me not to go for a run, some pictures popped up on my phone. There, peering out of the mist of drifting over-spray, was Peggy, resplendent in her newly applied coat of

Snowberry white. We had arrived at a place I thought we might never reach.

I visited the garage the next weekend and it was wonderful. Peggy was gleaming. There were a few bits of overspray and some areas lower on the car were a little more 'orange peel' than I might have wished, but it wasn't at all bad. We sprayed all the visible interior panels so it looked like it had been white from the day it rolled off the line in February '67. Some panels did need to be blown over again and this took a lot of thinners. At one stage, I really did get concerned that Barry was sneakily drinking the stuff. Although he did have a faint white ring around his mouth, I gave him the benefit of the doubt and considered that this must have been where his mask had been while he was spraying.

It wasn't the most healthy of environments. I am afraid that, at this stage, we were not going to get any awards for safety in the workplace, and the only effort made to dispel the fumes came from Barry's constant smoking. I did think at one point that he was contemplating making a hole in the mask for his cigarettes.

After much flatting and respraying, we arrived at a stage where neither of us had detectable fingerprints. Any spare or worn bits of underwear and T-shirts had been used in wiping Peggy down. Eventually, one Sunday, we opened the garage door and the light flooded in, pushing away the gloom of the garage, revealing Peggy in her bright new clothes. We stood like a pair of self-satisfied Cheshire Cats waiting to have their photo taken. It was a strangely emotional moment, partly because of our achievement and partly because Barry smiling was not a vision that any

Peggy in her bright new clothes.

person should have to observe for too long. We didn't say anything; we didn't need to. There was an almost spiritual silence. Then Barry scratched his backside and broke wind. We shut the garage door and headed home.

I sat in the pub that night thinking again of Roger Bannister. A strange thought came into my mind. What if Barry had been his pacemaker? The image came vividly to me; Roger Bannister emerging through a hail of fag smoke. I burst out laughing. The pub fell silent for thirty seconds and several regulars nudged each other and pointed in my direction. I didn't care, that night I was in possession of that most temporary of feelings – happiness.

16

# A Moment of Supreme Decision Making

Disappointment is something we all face in life, some of us more than others. My experience is that there are various levels of disappointment that can roughly tally with different types of electric shock. Some electric shocks make you jump backwards with a little pain and embarrassment, but you manage to pull yourself together and carry on. Others really hurt and, even though you try to hide it, you know that you dare not look at the part of your body that feels as if it has just melted. It takes time to get past disappointments at this level; lingering self-doubt and the feeling of stupidity at a self-inflicted setback remain in your mind. Of course, some electric shocks are sadly fatal. It's kind of a way of making sure you don't do it again, although I feel fairly sure that this isn't in any health and safety 'online' training advice.

My disappointment with Barry was hovering somewhere towards the end of the second stage. I feared the last stage was looming in the near future. I felt myself constantly at the 12-volt shock stage of disappointment. You will all know what 12 volts feel like. You don't put your hand back on the spark plug again immediately because you have just jerked your head upwards at a rate of knots and banged it off the underside of the bonnet. Yes, that's where I was at alright; shaken but not stirred.

The resurrection of Peggy was within touching distance but Barry had become even more elusive than usual. My trips to his garage had resulted in a series of no-shows that rivalled the concert career of the late Meatloaf. Indeed, Barry was more like the late *Malt*loaf. What had happened? I was concerned. Had Barry run off with Lauren, the woman of his dreams? My mind was filled with an image of Barry running along a beach in slow motion. Everything that Barry did was in slow motion, so it would not have required any extra special effects to capture this vision. I shook my head violently in a vain attempt to scrub the image from my tortured brain but the show continued. I now saw the pair of them running into the distance hand in hand with a welding rod around Lauren's waist. I saw Barry's wife, Kelly racing behind the eloping pair much in the manner of an inebriated space hopper, her fist clenched tightly on a can of Red Bull. I shook my head again to try and dispel this vision, lost my balance and fell over. A sympathetic lady nearby in the supermarket helped me back up but steered me very determinedly away from the drinks aisle. I thought about explaining the complex story of Barry and Peggy but realised very quickly that both she and the

Into the distance hand in hand.

security guard would not have understood. Indeed, it may have made matters worse.

I thought long and hard about what to do. Marilyn suggested finding someone else to help me, but that would mean relocating the car and paying for another garage. A few weeks passed and suddenly Barry phoned me out of the blue. Well, actually, out of the phone he had borrowed from a friend, as he had no credit left on his. He suggested we meet at the garage, but could I pick him up? I duly obliged and on the way over to the garage, Barry told me that his benefits had stopped so he had no money. And Kelly was pregnant! More importantly, though, he told me that he'd had a message from a mate who was selling a 1980s basic Mini and was I interested? In a moment of supreme decision making, I said, 'Yes!'

We went to look at the car and it was pretty rotten, but I knew Barry could put it back on the road. Two 'A' panels, a sill, a front nearside floor pan and a respray would see us quids in, I reckoned. Especially given the exorbitant price of vintage Minis. Now you may think that this was complete madness on my part, involving some less-than-focused decisions, but consider this; I needed to re-engage Barry so the Mini would provide me with a route in and extra money on top.

£500 secured the Mini and we were back in business. I explained the purchase of the Mini to Marilyn. The conversation could have been lifted from *Jack and the Beanstalk*, where the main character explains to his mother the exchange of the cow for some magic beans. Marilyn used one of her assorted expressions normally reserved for the problem cases in social work. I must admit, she hid her enthusiasm for the project well.

The Mini wound up at Barry's garage in another space he had next door, behind some bushes. Driving the Mini was reminiscent of the '60s *Batman* series, complete with the fire emerging from the exhaust of the vehicle.

Work soon started back on Peggy and next up was the electrics. We had pulled everything out of her and needless to say we had failed to keep any record of what pieces went where. We all know that over the course of many years, our classic cars were subject to the whims of previous owners. This often involves incorrectly wiring in a radio, adding unnecessary extra lights, creating extra switches that operate nothing, and so on. In the process, they introduced negative wiring to positive wiring, colour-coded wiring to household wiring, in-line fuses to bits of coat hanger. The list is, unfortunately, endless.

So, the next day found myself and Barry trying to match wiring. The wiring diagram was some help in defining what we should have found. However, Mr Haynes' handy manual did not cover the bafflingly twisted bird's nest of wires that we actually found. In the end, we located most stuff through trial and error. The indicators came to life, brake lights reflected satisfyingly off the rear of the garage, even the floor dip switch worked and the wipers clawed their way reluctantly across the front screen.

The only issue we had was the headlight flasher. It was operated from the indicator stalk with the green teardrop light at the end of it. The original one had disintegrated, so I had purchased one on eBay. When it arrived it looked in great nick and we quickly installed it. The indicators worked fine, I moved it forward for the headlight flash and … no forward movement. The indicator stalk had no flasher component and no extra set of terminals to accommodate the function. To this day, I have still not been able to source an indicator stalk that has the original set of functions.

We then turned our attention to the brake pipes and the plastic mock chrome bead around the rear tailgate screen. You may think it strange that I have lumped these two in together and, of course, in the sequence of any other restoration story you would be right. But remember, this story is a lesson not so much of success but of a cautionary tale outlining how to avoid success at every turn. The connection between the bead and the brake pipes is a simple one. They both took the most circuitous route to get to where they were needed. Barry had apparently fashioned the brake pipes following the trail of a nomadic snail that had traversed the underneath of the car. The bead on the tailgate looked as if it had been applied from a tube of toothpaste.

That night I parked the trusty Micra up and strolled down to the pub. As I absorbed more pints of 'Old and Filthy', a heavy cloak of disappointment grew over me. Four years into the project and all I had to show for it were wobbly brake pipes and a rusty Mini. On the way home, I noticed some wires sticking out of a nearby street light. I wondered, should I try and match them up? Better not push my luck.

What have I done? One week after Peggy's extraction from the farmer's field.

In a rented garage in Deal, Kent. Note the roll of vinyl bought to re-cover the seats.

Peggy illegally parked on Deal seafront.

Peggy meets a Ford Cortina MK1 on the way to the chip shop on Deal seafront.

Peggy and pals! Peggy lines up alongside a Ford Escort MK2, a Ford Sierra RS Cosworth and a Ford Anglia.

Peggy and her good friend Morris.

Peggy meets a VW Campervan.

Harry with Peggy at a car wash fundraiser for the RNLI.

Peggy flouts the law again on the double yellows.

Peggy stops for a gallon of regular at Dover Transport Museum.

Two icons! East meets West. Peggy with a Trabant at East Kent Railway Trust, Dover.

Peggy phones it in at Shepherdswell, Kent.

Home sweet home. Peggy is happy in her current home in Deal, Kent.

Harry and Marilyn on Deal Pier.

17

# A Silly Man and his Silly Dream About a Silly Car

In my experience of planning and project work, advice would usually centre around topics such as stage analysis, critical pathways and contingency planning. There is no doubt that although most of us do try and set goals, our personal ventures are born out of love and a modicum of over-belief in our own abilities. The restoration of our cars certainly falls under the definition of 'project'. They are projects so bedevilled by unforeseen, or 'should have been foreseen', issues that our initial timelines have often expanded beyond recognition. It is as if they had wandered too close to some undocumented black hole or gateway to another dimension.

With this in mind, it was with some comfort that I learned of the aborted Artemis missions to the Moon. The poor folks were just ready to go when they found some sort of a leak. I can see it now, accusing eyes scanning around the mission control room. The question on everybody's lips: 'Who was supposed to tighten that bit up, eh?' Somebody chirps in with: 'Well, I don't want to talk out of turn but the torque wrench wasn't left in its proper place …' This is soon followed by: 'Yeah, I saw a couple of nylon washers lying around last week.'

Well, the American Space Agency are lucky. They have a team of top brains dedicated to their mission and they will, of course, find a way. I had Barry, who was dedicated to being missing and who always found a way to be somewhere else.

It was now three-quarters of the way through 2013. The project had overrun – even Doctor Who would have had to acknowledge that as a fact. Things crawled on at a snail's pace through Christmas that year. We did add a new radiator and we put most of the interior back in, albeit in a very rudimentary way. There was no carpet, just seats on a bitumen-black floor full of welding seams. The door rubbers were acquired from a local scrapyard. Although from a modern vehicle, they worked remarkably well. As 2014 slipped past, I realised the one control point in the project I'd missed was time. I just hadn't factored in my inability to alter time. It was a schoolboy error.

However, I do recall the first ride in Peggy. One Sunday afternoon at about 15:30, watched by Barry (who looked more than a little nervous), I turned the key on the rust-scored metal dash and Peggy coughed into life. Then slowly, like a person who had been held captive in a dark cell for years, Peggy emerged blinking into a day shrouded by the grey and overcast April sky. There was no fanfare, there was no announcement, no speech by the local mayor.

There was no fanfare.

There was no champagne bottle that might have been fittingly smashed over the head of the driver. Nothing but me, Peggy and Barry.

We shut the garage door down and Barry climbed in. Off I went, gingerly around the block, faltering at first. In my mind were two thoughts. This was the first time that Peggy had moved under her own efforts since her last drive into the back of the farmer's field in 2004, where she had been left to her fate amongst the cattle, the local children and the grass that grew through holes in the floor pans. Her rescue now seemed as though it had been affected by someone else and not by me. It seemed like a story I had read somewhere about a silly man and his silly dream about a silly car. I noticed that the windscreen had started to mist, or was it my eyes?

The second thought I had was that we were travelling along in something that had been put back together by me and Barry. I realised it was now too late to warn other road users.

We pulled Peggy back into Barry's garage, got out and looked at each other. The scene was very reminiscent of Humphrey Bogart and Claude Rains at the end of *Casablanca*. OK, minus Ingrid Bergman and an airport but hey, stay with me on this. I knew, unlike the last line of that film, that this was not the beginning of a beautiful friendship. It was not even the beginning of the end of one. It is true that what we had achieved together was, in its own small way, 'quite something'. And, indeed, it was something that we would both have in common even if we didn't have Paris. Instead, we had Norbury just outside Croydon and dirty hands and grubby fingernails. A week later, Peggy had a successful MOT – news I received while sitting at a child protection desk awaiting the kind of phone calls that you never get used to. It helped me to get through the day. I was a Cheshire Cat on the inside.

Change was afoot on many levels in 2014. I changed jobs and we made an offer on a house in Deal, Kent – a place we had fallen in love with by 'going on holiday by mistake' in 2006 (see *Withnail and I*). Our offer on the house was accepted and plans were made to move down in June that year. I informed Marilyn that, as we had downsized, I was perfectly capable of organising the move to Deal, which was no more than 70 miles down the M20. A Luton van was duly ordered from our local hire company. Even when ordering the van, I had a sense of unease about the guy behind the counter, who tried to reassure me that their only long wheel-based Luton van would be available on that Saturday.

As it turned out, their only long wheel-based Luton van was not available that Saturday and I trundled up to our house in a high-top transit that would only have accommodated our belongings if we had adapted them with a chainsaw. What to do? Barry? Yes, Barry! Barry was called and he duly helped us move. He was, of course, handsomely paid, but it gave us a chance to talk about the need to relocate Peggy to Deal. A plan was formulated. Once I had located a garage for Peggy, Barry would follow me down from Croydon to Deal. All I had to do was find a garage in Deal. What could be more simple? I sat in my new local after we had moved our furniture in and thought through the mission to bring Peggy down. 'It's not rocket science,' I thought. 'All I need is a plan …'

18

# Let the Wacky Races Begin!

So, the day came when Peggy once more had to relocate. She had started off in a large factory in the shining, swinging '60s. She had been a proud beacon of a brave new world where innovation and different thinking were going to sweep away the trammelled straitjacket of the past. Sadly, like the '60s, Peggy had burned brightly for a time and promised much but was all too quickly discarded. It always seems to me quite remarkable how swiftly that which had seemed unquestionably modern becomes a fading signifier of the way we once were. How futile we now know it all to have been. We allowed ourselves to dream and now those dreams have succumbed to a hard-headed future that finally caught up with us. We allowed that future to overtake us and our dreams had now become a reflected shadow that flickered in our rearview mirror.

And so it was, in the dim and distant past, Peggy went to live in Bradford-on-Avon, outgrew her usefulness and lay for six years buried up to her doors in a farmer's field. The journey back from that field is one you have read and I hope found common to your own experiences in both laughter and tears. For all the futility of life, we have achieved something, and our cars testify to this. It was something for ourselves but something for others also.

The journey from Norbury in Croydon to Deal in Kent really should not have held much fear for someone who had spent nearly five years restoring the vehicle set to undertake this journey. Indeed, Peggy was originally designed to very easily accommodate such journeys. But that was very nearly forty-eight years previously. In that time, Farmer Grunge had tried to bury her, children had jumped all over her roof, and rats had scuttled in and out of her, clinging to the foliage that had grown through the holes in her floor.

Both myself and Barry had thrown ourselves into battle against all of this in a bid to wrestle Peggy back to life. It was a very uneven contest. In five years of this undertaking, Peggy had robbed me of most of my savings and my sobriety. Barry had taken care of the rest. Barry nonchalantly took my wages, my belief in humanity, and my sanity and gifted me the pity of others, who shook their heads and often talked to my partner Marilyn in hushed tones while pointing at me when they thought I wasn't paying attention. Still, I was here and now we were to set off. Let the Wacky Races begin!

I had asked Barry to follow me down to Deal, travelling on 'A' roads. I had thought it a bit too cavalier to try and travel on the motorway. Of course, I could tell you that the journey was planned meticulously to avoid all possible challenges and outcomes. I could also tell you that the Moon is made of cheese but unless you'd envisaged me and Barry as Wallace and Gromit, we'd best leave both of those suppositions behind.

Barry could only make the journey at noon. This is a most ill-advised time to weave your way through Croydon, Penge and Bromley out to the A20. Many of you will be familiar with the 1994 film *Speed* with Sandra Bullock driving a bus. The gist of the film is that there is a bomb planted on the bus that would go off if the vehicle were to drop below 50mph. My problem was much of the other end of the speedo strip. I feared the outcome of going anywhere near 50.

In my mind's eye, I recalled every nut and bolt that Barry and I had undone and tightened on Peggy. More concerning, however, were all the nuts and bolts I had never got to see Barry tighten. Every creak and groan (and there were many) had me pressing on the brake and backing it up with the handbrake. The SAS would have done well to put their brightest prospects through such an ordeal in basic training. There would have been little doubt about their suitability for the most senseless of ventures if they had come through that with dry underwear. There was no sound damping in Peggy, no carpet, just the seats loosely fitted. At times, I felt like a 'B' movie actor in a '50s British film about breaking the sound barrier. All I needed was a 'Biggles' helmet or a space suit with Apollo 13 stitched onto the shoulder badge.

At Ashford, I took a wrong turn and wound up on the M20 and Barry followed. I could see his rotund face in

Into Deal without any mishap.

the rearview mirror. He sat in comfort, in his shiny new van, barely aware that I was wrestling desperately with a machine he had helped create. It was at this point, I thought, 'Sod it'. I just need to get to Deal. So, I put my foot down and, rattling along at a steady 60–65, we ploughed on to Deal via the M20 and yes, I even overtook someone! Coming over the hill from Folkestone to see Dover Harbour in all of its glory is always an impressive sight. On that Saturday, it was even more so. The light was fading faster than my powers of concentration, as we weaved our way down the last leg of the A258 into Deal without any mishap.

During the previous week, I had managed to hire a garage from an antique dealer. How very apt, I thought. The antique dealer charged me just something short of what it would cost to stay at the Savoy for a month but I was in no position to argue. I packed Peggy inside the garage and went over to speak to Barry. He didn't actually emerge from his shiny new vehicle. Peggy's arrival in Deal hadn't warranted that level of excitement. He leaned out of the window and said something like: 'It's done.' In truth, I couldn't hear what he said as the noise inside Peggy had reached the levels approaching a Motörhead concert. But I saw his lips move and I knew he was heading off. I paid him, and in one movement, the window squeaked up and he roared away just as I was saying, 'Keep in touch about the Mini.'

I watched the tail lights of Barry's van disappear back up the A258, melting into the dusk and becoming part of everything that was the distance. I still had business with Barry as we had the unfinished Mini project to sort. I realised then that he had left me still some 3 miles from home, but it was a nice evening and during the walk, I thought about all that we had shared. I thought about it long and hard. If I had the chance to turn back at that farmer's field in 2010, would I? If I could have ignored the bid button on eBay in 2010, would I? I wasn't awfully sure, knowing what I now knew, that I might not have made a different choice. I had met Barry, Lauren, and Kelly and survived them all. I was pretty certain I would never see them again and I was equally certain that if that was the case, I would probably have a better life expectancy.

I arrived back at our new house and cajoled Marilyn into going for a drink at our new local. I think she was quite surprised that I had made it back. I had a few pints and laughed. I laughed the kind of laughter often used in trailers for Hammer Horror films. People looked around. I tried to explain about Peggy but Marilyn took me by the hand and led me home, briefly explaining to the landlord, who then gave a knowing nod to the locals.

# 19

# I Found Myself Smiling

There are many wonders in this scientific world that I struggle to understand. I once asked an astrophysicist about the theory of the expanding universe. After an enthralling lecture at the Greenwich Planetarium, I held my hand up for questions afterwards. No one else had done this and I had a good idea why. In these situations, there is a quite natural fear that the question you ask will promote uproarious laughter from all in the auditorium while the Professor, in shock at the stupidity of your question, stumbles backwards, spilling his glass of water onto the display control unit. Sparks, followed by blue flames would then ensue, causing panic in the auditorium as thick black smoke quickly filled the space left by adults and children vaulting over the seats in some crazy re-enactment of a nineteenth-century steeple chase.

Of course, you will be comforted by the knowledge that it didn't quite happen like that. The question I asked was this: If (according to Einstein et al) the universe was expanding, then what was it expanding into? The Prof was dumbfounded and quickly mumbled something about the lecture having overrun and being late for a train. I knew this was the likely scenario before I asked the question. By the way, if you *do know* what the universe is expanding into, can you please let me know via my private email and we can perhaps share the Nobel Prize money at a dinner time session in the Dog and Duck. Following the question, I don't mind admitting that I had felt a little smug. In my experience, it doesn't matter how smart you think you are, someone will always ask that childlike question that completely floors you.

Since I was no longer living anywhere near Barry, my questions to him had travelled at first via texts. They were less along the lines of what the universe was expanding into and more directed at what his level of inactivity and lack of response were expanding into. Although Peggy was now officially back on the road, my ties with Barry remained unbroken due to 'PROJECT MINI'. You will recall that in a moment of inspired madness and utter insanity, I had invested a lot of money in purchasing a very rotten Mini along with sills, 'A' panels and wings. Of course, this was not an investment on the scale of Elon Musk or indeed HS2 but, nevertheless, it represented a severe dent in my beer funds. Looking back, I now realise that this decision could be compared to Captain Smith of the *Titanic* striding on deck and ordering his officers to actively find an iceberg to steer towards.

After no understandable text responses, I tried phoning Barry on many occasions with no success. In the end, I gave up. Being angry can seriously upset you and I had no wish to be continually angry. Marilyn counselled that I had something that I would never have achieved without my interaction with Barry and perhaps £1,000 worth of rotten Mini was a reasonable handshake to leave him with. Marilyn was right on some of the issues, but Barry had also helped me achieve occasional levels of poverty and alcohol dependency that I would have struggled to achieve on my own. But I did have Peggy.

Marilyn counselled.

I never saw Barry again and, in truth, I had no great wish to. I had a notion that this feeling was mutual. What is he doing now? Who can tell? Chasing Lauren, being chased by his wife, or is he back in the middle of the children's fairground ride in the centre of Bromley High Street, where I first found him? If you are ever in Bromley High Street and you see a chap who looks a little like a cross between James Corden and Oliver Hardy, struggling with excited children and demanding moms in the middle of a miniature fairground ride, take a photo and send it to me.

As for Peggy, I had to move from the garage I originally rented in Deal when I was roundly abused by a very posh woman who objected to my 'continual tinkering'. I was lucky enough, sometime later, to get a garage with the local council. With a modicum of trepidation, I took Peggy to a local garage in Sandwich that specialises in classic cars. I thought that it would be a good idea to get an MOT that didn't involve people that Barry knew. Peggy mostly passed but some of Barry's welding was highlighted. Where we had let in the floor pans, the join between the central tunnel and the floor had been welded from above and not below. A seam was re-welded from below and all was well. I did have some problems with the fuse box and I found that the indicators intermittently dropped out of function. You may know that the fuse box on our cars was commissioned only just before the discovery of electricity and, consequently, is not the best. However, on inspection, the problem turned out to be the insertion of a 3/16 bolt instead of a fuse. 'Good old Barry!' I thought and despite it all, I found myself smiling, a habit I try to keep to a minimum.

So, late one night in the Ship Inn after being in Deal for twelve months, I reflected on Peggy and Barry. If the universe was still expanding, it would explain why the satellites tasked with conveying my texts to Barry were not able to field any replies. Barry was an expanded 130 miles further away than he had been and, although the Earth was revolving around the Sun, a small carousel was revolving around Barry. Such infinite wonders beyond our comprehension, I thought ...

20

# The Work is Never Over

I find it very instructive that when we talk about our cars, we use the word *restoring* and not *restored*. 'I am restoring an Austin 1100', as opposed to 'I have restored an Austin 1100'. I am quite sure that many members of classic car clubs would be able to tell me how many singular components there are to an Austin 1100 and, yes, you lot – it doesn't matter how well you have 'restored' your car, for all the wonderfully inspiring examples you show each week, chances are you will not have replaced everything. Every single original component is over 50 years old right now, down to the accelerator return spring. Replaced that, have we?

We all know the work is never over. That realisation either defeats you or draws you on. Think of all the 'Friday' or 'Monday morning' components still in your car. Think of all the people who owned the vehicle before you. These were people who had shares in fibreglass resin and pop rivet guns and were addicted to the smell of Isopon. You might look under your car and see that the wording on the oil filter is in Latin and then you notice the phrase written on the end of your misshapen exhaust tailpipe is '57 varieties'. And there's more. There are the wires that you trace twice around your car until you find they go nowhere and do nothing. That is until you cut that wire, and then for weeks afterwards, your wipers don't work or your indicators come on every time you press the horn.

During the Peggy story, I will have told some tales and relived some experiences that many of you will be familiar with. Of course, you didn't have the added ingredient of Barry. And no, I don't miss him! I still find little bits that have Barry's signature all over them. I often wonder what he is doing now. I don't do this for too long as it makes me a little uneasy and leaves me with mind pictures that would require a spark plug brush to remove. Has he gone back to sitting in the middle of that kids' merry-go-round in Bromley High Street? Apparently not. Marilyn went back to Bromley recently and informed me that there was a merry-go-round but there was no man sitting in the middle, just a kindly lady helping the kids on board. I wondered if the kindly lady was any good at welding. Marilyn said she would ask next time she was up that way.

I, of course, have needed to do running repairs on Peggy since we arrived in Deal in 2014. I now have a garage in my driveway, which has conferred two main benefits:

- I have electricity and this enables me to use power tools.
- People in the quiet streets of Deal cannot hear me swear.

I have recently learned to weld with results of varying quality. Marilyn described the scene in my garage as one where she expected a fully formed version of Boris Karloff as Frankenstein's Monster to emerge from amongst the sparks and clouds of smoke. Instead, it was worse than that – just me with a welding helmet on, staggering around the yard trying to unclip it from my head with my burning gloves. I recently replaced a door on Peggy by drilling out the bolts and re-tapping the holes to fit new ones. One day I will look back and laugh, but not just yet.

Peggy often goes to work with me at Dover Council, much to the bemusement of locals. Peggy also goes on Sunday trips to Folkestone, Ramsgate and Canterbury. During those trips, sometimes people stop me and ask, 'How did you get hold of that car?' I reel myself in from explaining, as I realise that they would never understand. But you, loyal readers, continually in the process of restoring, you would very much understand.

At work we have a young apprentice called Tom. He is young, enthusiastic, clever and very naive. He reminds me of myself all those years ago, if you remove the clever bit. Some days I sit in Peggy in the garage and I am whisked back to 1976 with OOG 300G, remembering driving lessons from my Dad that included the whole family in the back. I had three brothers continually fighting and jostling for position, with my Mom in between acting like a boxing referee. Issigonis and Moulton never bargained that they had invented a time machine.

Thank you all so much for tuning into my story. It has been a massively self-indulgent honour to share the tale with you all. At the end of the day, it has been the tales

that other classic car restorers have told of their ingenuity, perseverance, skill and dedication that have been truly inspirational. Most impressive of all has been the willingness of the Austin 1100 Club community to share, advise and help. These are qualities the Club has that are just as valuable as any one of our vehicles.

Anyway, I must crack on. Where's that accelerator spring?

Many thanks, from Peggy and Harry.

# Getting There ...

In June 2023, around two months after completing his last Peggy Restoration article for the club magazine, Harry embarked on one of the most important journeys of his life – a journey of rehabilitation and recovery. While running along the seafront in the East Kent coastal town of Deal, 'keeping fit' ironically, he suffered a life-changing, catastrophic haemorrhagic stroke. It was a left-brain bleed, leaving him with right-side paralysis and damage to the basal ganglia area of his brain, severely affecting his speech.

Since then, Harry has adopted the same doggedly determined, bloody-minded approach as he did to restoring his beloved Peggy, but this time, instead of rewiring the car radio, he is working hard to rewire his brain. It turns out that although the human brain is the most complex entity in the known universe, 100 billion neurons strong, some of the best tools for rewiring it are very simple and involve focused and dedicated practice. Most important of all is repetition: repetition of movement and tasks, of learning to speak and beginning to walk again. Neuroscientist Sam Wang described the brain as less like a computer and more like a very busy Chinese restaurant – chaos, but everything gets done.

Although he is making significant progress, with potential to achieve more, Harry can no longer drive Peggy around. Luckily, through his local community work and having played in three bands since living in Deal over the past eleven years, he has built up an admirable network of good friends. Among them is Pete, who takes Harry and Peggy out for a regular spin, much to the delight of passersmmmmby. Pete's still trying to come to terms with Peggy's gearbox, likening it to a Rubik's Cube puzzle, proving that there is yet more work to be done on the Austin 1100 Estate.

There is also more work to be done on Harry, and he's more than up for it. He is currently working with a team of specialist neurological therapists to maximize his recovery potential. We can't pretend this is easy. It's a well-worn cliché, but it really is a roller coaster of emotions, pain and bloody hard work, just as with Peggy! It's like suddenly being told you now have to learn French, learn gymnastics and, oh yes, learn to play the piano, all at the same time. Not to mention dealing with the emotions and sudden loss.

It's not all doom and gloom though – far from it. Harry's now in a new band, Last Man Standing (no pun intended). They play all original songs. He plays the keyboards with one hand, arranges the songs and provides instrumental support to the band's guitars and vocals. In the words of one of the other band members, he 'holds the band together during performances'. He's also hoping to get back under the bonnet of the 1100 Estate one day, and he is working hard to achieve that aim.

He'll get there, we know he will, just like Peggy did from the farmer's field all those years ago.

Marilyn

The band! Last Man Standing – left to right: Nick Holbrook-Sutcliffe, Harry Taylor, Sharline Primrose, Johnny Jahlfrezi Morris and Kim Nash. (Kay Valentine)

# Afterword

A few weeks ago a dear friend of mine died, and just before writing this I was standing at the spot where her ashes were to be interred, planning what I would plant there; something with some staying power, a bit of wild thyme, a few oxlips, a nodding violet, a little china rose. A few yards away I passed another grave – Harry Whetter, Churchwarden of this Parish and his wife Margaret, known to all as Peggy. I stood there for a few moments contemplating the forty one years I've lived in this beautiful village and how fast the years had flown.

So what's that got to do with restoring cars? Well, I was secretary in turn for many years to the engineers responsible for designing and developing the ADO16 – Charles Griffin and Alec Issigonis. I was very young to begin with, which is how I'm still around to tell the tale. Charles, who was a brilliant development engineer, absolutely loved the car and said if he never did anything else ever again he would be proud to call it the fulfilment of his life's work. Sir Alec, on the other hand, quickly moved on: his heart belonged to his next project, which was the one never to be built – the Morris Minor, the Mini and the 1100 were his past affairs. You win some and you lose some, that's just life. 'Build me up Buttercup, don't break my heart' was the tune of the moment, but the failure of 9X made him wish he'd gone into aircraft instead of motorcars.

I also heard on the BBC News not so long ago an announcement about the date when the stars first began to shine, after the Big Bang; it was 250–380 million years ago that the Cosmic Dark ages gave way to the Cosmic Dawn and there was light. That discovery was the Holy Grail to the astronomers who had pursued it. The emerging evidence was a dream come true; the stuff that exploded is the stuff we are made of, stardust. And I must say it blew me away! Now that's pretty much the story of *When Harry Met Peggy* and what can happen when you meet up with another lump of stardust; it's a story of destiny and resurrection in all its wonderful mystery. And I'm sure they'd smile at Harry's great endeavour, Alec and Charles. Yes, in the beginning we created the BMC 1100 and it was good; it was something more than just another car, wasn't it.

Suzanne Johns
12 October 2025

# Glossary

**Foreword**

Hydrolastic
A fluid-based suspension system.

**Prologue: 1967**

*Carry On*
A saucy British comedy film franchise, comprising thirty-one films.

GCHQ
Government Communications Headquarters.

Flower Power
A hippy movement in the late 1960s.

Donovan
Donovan Phillips Leitch, a Scottish musician, popular during the late 1960s.

Brean
A popular seaside resort village in Somerset, England.

*Now That's What I Call Music*
A series of compilation albums started in the 1980s.

Frank Capra
An Italian American film director who made *It's a Wonderful Life*. It's the story of a man who puts the needs of others before his own. In a moment of crisis, he is visited by an angel called Clarence, who shows him how different the world would have been without his contribution.

## 1 Snowberry White, with a Black Dual Strip

Carburettor
A mechanical device that mixes air and fuel to create the combustible mixture for petrol in an internal combustion engine.

Gearbox
This contains a gear set. There are normally four or five forward gears and one reverse gear.

Rod change
An improvement on the previous flexible gear change parts, using solid rods and joints.

## 2 How Hard Could it Be?

Thomas Hardy
English novelist and poet. He set his novels in the rustic and rural landscape of southern England.

*The Twilight Zone*
An American sci-fi television series featuring Rod Sterling. It was well known for its bizarre and unexpected storylines.

*Withnail and I*
A black comedy film written and directed by Bruce Robinson, made in 1987 and set in 1969.

## 3 Almost Without Realising, I Had Started to Care

Floor pan
A large sheet of stamped metal forming the foundation of the floor.

Inner wings
These are metal panels just beneath the outer wings and form the engine bay.

Sills
Also known as rocker panels, these are the structural components running along the bottom of the vehicle.

Exhaust
Used to guide gases away from the controlled combustion inside an engine.

*The Killing Fields*
A 1984 film about the Cambodian War. It features Barber's 'Adagio for Strings', originally written in 1938 with help from Arturo Toscanini.

## 4 This Could Be the Start of a Beautiful Friendship

Axle stands
These are used to hold the vehicle up when it is not on its wheels.

Two-tonne trolley jack
A heavy-duty floor jack to lift the vehicle.

Scissor jack
A portable jack used for changing a car wheel.

Subframe
A structural steel component that bolts to the vehicle's chassis.

Christopher Lee
A film actor best remembered for his portrayal of Dracula.

Pressed Steel Fisher
A company that manufactures car body parts.

Captain Smith
The captain of the ill-fated *Titanic* that sank on her maiden
voyage in 1912.

Thomas Andrews
The designer of the *Titanic*. He was aboard the *Titanic* when
she went down.

Howard Carter
British archaeologist who discovered the tomb of
Tutankhamun.

Valerie Singleton
Presenter of the BBC children's television series *Blue Peter*
from 1962 to 1972.

Tony Hart
English artist and children's television presenter, best known
for his plasticine character, Morph.

Ray Charles
American singer-songwriter who was blind.

Stevie Wonder
American singer-songwriter who was also blind.

Peter Lorre
An American actor known for his timidly devious characters.

Humphrey Bogart
In the film *Casablanca*, Bogart plays Rick, who utters the
immortal words: 'This could be the beginning of a beautiful
friendship.'


## 5 The Smell of Melted Car Radio

Banks crashing
There was a disastrous financial crisis, causing banks to fail
after 2007. It required a substantial government bailout. A very
frightening time for us all!

Alfred Hitchcock
An English film director. He is widely regarded as one of the most
influential figures in the history of cinema. His film *Strangers on
a Train* featured a thrilling scene set on a runaway carousel.

G Force
A measurement of acceleration, usually associated with test pilots and astronauts.

Van de Graaff generator
An electrostatic generator, creating a very high voltage.

Apollo astronauts
The Apollo programme flew nine missions to the Moon between 1968 and 1972.

Billy the Train
A bright orange train, one of Thomas the Tank Engine's friends.

Doppler effect
The change in pitch as a moving source approaches our ear.

Fan blade
One of the specially shaped rotating parts of a mechanical fan.

Water pump pulley
Part of the vehicle that keeps the engine from overheating.

Choke
A valve in the carburettor that is used to reduce the amount of air in the fuel mixture.

Sump
A depression or reservoir designed to collect liquids.

Spark plugs
A device for delivering electric current from an ignition system.

Earth
This allows for safe dissipation of electrical charges.

*Old and Filthy*
Refers to a 1969 comedy recording by the British actor Ronnie Barker.

Cheshire Cat
A fictional cat famous for its wide grin, popularised by Lewis Carroll in his novel *Alice in Wonderland*.

**6 In a Galaxy Far, Far Away**

Tony Hancock
An English comedian who starred in *The Rebel*. In the film, the character of Mrs Cravat was played by Irene Handl.

Baron Von Frankenstein
A character from Mary Shelley's book *Frankenstein*. The mad baron was played by Peter Cushing in a film version from 1957.

Float chamber
A device for automatically regulating the supply of a liquid to a system.

Electric pump
Moves fuel from the tank.

Temperature gauge
Monitors the engine coolant temperature.

Clutch
A mechanical device that disconnects the engine from the transmission, allowing the driver to change gears.

Wormhole
A theoretical tunnel in space that connects distant locations.

ADO16
The Austin 1100 and 1300 were often called ADO16 by car enthusiasts.

*Oliver Twist*
A novel about a young orphan, by Charles Dickens.

Moon Base Alpha
The setting of the British science fiction TV series *Space 1999*.


**7 She'll Have to Go**

Laurel and Hardy
An early American cinematic comic duo, par excellence.

Model 'T'
An automobile that was produced by Ford Motor Company
from 1908 to 1927.

Van Morrison
A Northern Irish singer-songwriter. Well known for turning his
back on his audience.

Ferrous metal
A category of metal primarily composed of iron.

Lower radius curve
The smallest curve achievable. It depends on the thickness of
the material. This one was perfection.

Francis Drake
Drake was in command of the English fleet that defeated the
Spanish Armada in 1588, after a little game of bowls!

Dulux Weathershield
A weatherproof paint.


## 8 A Game of Vehicular Russian Roulette

Wrong handed car part
Receiving a part for the wrong side of the car.

Apollo 11 and 13
Apollo 11 was the first manned space rocket to go to the Moon.
Apollo 13 was to be the third lunar landing attempt, but the
mission was aborted after the rupture of a service module
oxygen tank.

Tow arm
A device used to tow another vehicle.

Stanley Unwin
A British comic actor who invented his own nonsensical
language called 'Unwinese'.

George Orwell
English novelist and poet, author of *1984.*

Winston Smith
The main protagonist of *1984*.

Brake and clutch cylinder
The brake pedal stops the car while the clutch pedal disengages
the engine. Both contain a small cylinder.

Heater box
The box under the dashboard that generates heat.

Bulkhead
In a vehicle, the bulkhead is a structural partition.


## 9 Another Attempt to Defy the Laws of Physics

Alexei Navalny
A Russian lawyer, activist and political prisoner until his
untimely death.

Vanden Plas
A very posh version of an 1100 motor car.

Quarter light
A small side window in a car.

Displacers
Rubber and fluid parts that help the car ride smoothly and stay
level over bumps and around corners.

The Gloucester Marbles
The name I have given to the car parts I obtained from
Gloucester, like the Elgin Marbles, but smaller and more
controversial. (Gloucester does not want them back!)


## 10 The Lights Go On and So Does the Kettle

Confucius
A Chinese philosopher who died in 479 BCE.

Dalai Lama
Spiritual leader of Tibetan Buddhists.

Cross ply and radial
Two different types of tyres. You should not mix them on the
same vehicle.

Subframe towers
Where the suspension displacers are housed.

Engine head
The component that sits on top of the engine block.

Sherlock Holmes
The great fictional detective created by Sir Arthur Conan Doyle
in 1887.

PG Tips
A brand of tea famously advertised back in the day by using a
group of chimpanzees holding a tea party.

Johnny Morris
A much-loved presenter for children's TV who created a
programme called *Animal Magic* and provided amusing voices
for the animals.


## 11 Where Was Charlton Heston When I Needed Him?

Charlton Heston, James Mason and Elizabeth Taylor
Film stars from the 1950s, during the Golden Age of
Hollywood.

*Goodfellas*
A 1990 movie by the American filmmaker Martin Scorsese
about the rise and fall of Mafia hitmen.

Polarity
The direction of the electric current in a car.

Engine mounts
The rubber and metal fittings that hold the engine securely in
place.

Dynamo
A device used in motor vehicles to generate electricity.

Archibald McIndoe
A plastic surgeon who pioneered the treatment of Second
World War pilots who were horrifically burned and injured
during the conflict.

## 12 Looking Down the Wrong End of the Telescope

Roller kit
A set of components used to install a hydrolastic suspension
system.

Ben Gunn and Long John Silver
Characters from Robert Louis Stevenson's 1883 novel *Treasure
Island*.

Sid James
A South African–British actor famous for his raucous laugh
and his roles in the *Carry On* films.

Bushes
Bushes on a car are rubber or polyurethane and are fitted
between metal parts to cushion and reduce friction.

Grease gun pump
It acts as a hydrolastic pump for the suspension.

Plumber Boss White
A multi-purpose jointing compound.

Machiavellian
From Italian philosopher Niccolo Machiavelli: a term used to
describe someone who is cunning and lacking a moral code.

## 13 Whodunit?

Front ball joint
A greaseable joint that allows the wheel to turn and move with
the suspension.

Peter Cushing
A British actor best known for his roles in horror films.

Agatha Christie
English author known for her whodunit crime novels.

## 14 What Next?

Houdini
American illusionist and escape artist.

Yuri Gagarin
Soviet cosmonaut who, in 1961, was the first man in space.

Quarter panels
The body panel of a car between the rear door and the boot.

Wheel arches
The semicircular part positioned above the wheel of a vehicle.

Gielgud Theatre
Named after the English actor and theatre director, Sir John
Gielgud.

Dalek
The Daleks are a fictional race of extremely xenophobic
mutants portrayed in *Doctor Who*. They used to have a problem
with stairs.

Yoko Ono
Avant-garde artist, singer and songwriter who was married to
John Lennon.

## 15 The Fatal Lure of the Heroic Failure

Roger Bannister
A middle-distance runner who, in 1954, ran the first sub-four-
minute mile.

Overspray
The application of paint onto an unintended location.

## 16 A Moment of Supreme Decision Making

Meatloaf
American singer and actor, famous for the album *Bat Out of Hell*.

'A' panels
Panels fitted to a Mini to help give the car shape and support.

Haynes Manual
Workshop manuals for maintaining and repairing any type of car.

Brake pipes
Tubes that transfer brake fluid. They are crucial for the safe running of a vehicle.


## 17 A Silly Man and his Silly Dream about a Silly Car

Artemis Mission
The NASA programme intended to land a man on the Moon after 1972.

Claude Rains
A British actor to whom Humphrey Bogart delivered the line 'This could be the beginning of a beautiful friendship' in the film *Casablanca*.


## 18 Let the Wacky Races Begin

*Wacky Races*
An American cartoon series about a crazy car race.

*Wallace and Gromit*
Wallace, a cheese-loving bachelor, and his dog, Gromit, were the main characters in a British animated series.

*Speed*
A film about a disturbed, dangerous man who plants a bomb on a bus that is set to detonate should the speed drop below 50mph.

SAS
An elite military unit whose motto is Who Dares Wins.

Biggles
A fictional heroic pilot and adventurer in several books written for children by W.E. Johns.

Einstein
A German-born theoretical physicist known for his Theory of Relativity.

Motörhead
A British heavy rock band.

Hammer Horror
A British film production company best known for a series of lurid, Gothic horror films.

## 19 I found Myself Smiling

Elon Musk
Apparently, the world's richest bloke!

## 20 The Work is Never Over

Heinz 57 Varieties
A famous marketing slogan for the H.J. Heinz food company.

Spark plug brush
A wooden handle with bristles for cleaning spark plugs.

Boris Karloff
British actor who played Frankenstein's Monster in 1931.

Re-tapping
The process of running a tap (a specialised tool) through an existing, lightly damaged hole to restore its internal threads.